THIRD EDITION

Science Experiments and Projects for Students

Authors in alphabetical order:

Julia H. Cothron
Mathematics and Science Center
Richmond, VA

Ronald N. Giese
The College of William and Mary
Williamsburg, VA

Richard J. Rezba
Virginia Commonwealth University
Richmond, VA

KENDALL/HUNT PUBLISHING COMPANY
4050 Westmark Drive Dubuque, Iowa 52002

Disclaimer

Adult supervision is required when working on projects. Use proper equipment (gloves, forceps, safety glasses, etc.) and take other safety precautions such as tying up loose hair and clothing and washing your hands when the work is done. Use extra care with chemicals, dry ice, boiling water, or any heating elements. Hazardous chemicals and live cultures (organisms) must be handled and disposed of according to appropriate directions from your adult advisor. Follow your science fair's rules and regulations and the standard scientific practices and procedures required by your school. No responsibility is implied or taken for anyone who sustains injuries as a result of using the materials or ideas, or performing the procedures described in this book.

Additional safety precautions and warnings are mentioned throughout the text and in the front matter. If you use common sense and make safety a first consideration, you will create a safe, fun, educational, and rewarding project.

ISBN 0-7872-6478-4

Printed in the United States of America
10 9 8 7 6 5 4 3 2 1

Contents

Introduction

Dear Student:

In any sport you must learn to react correctly, quickly, and almost instinctively. The quality of your reactions can lead to your being named 'Most Valuable Player' or to your being quickly eliminated from the competition. Sports are interactive, and so is this book. You will be asked to take actions—to do activities, collect data, and answer questions. As in any sport, if you ignore the calls to practice, you do so at your own peril. It is important that you stop, do the activities, collect the data, and answer the questions as they occur. Practice will improve your skill level. You will need highly developed skills to be ready for the ultimate challenge when it is presented.

In many sports the terms leagues, conferences, brackets, and divisions describe different levels of sophistication of play. This book also has several levels of sophistication. In the first level you will learn each idea and skill in a structured activity. Then, you will be asked to do several practice problems. If you skip over the activities, questions, or practice problems, your skill level will fail you when you reach the 'world class level' activities—the design of original experiments.

At the world class level, you must understand the ideas and the skills so well that you can apply them instinctively with ease. By practicing you build your understanding. Knowing terms and their definitions are important, but not as important as being able to apply them. No one is perfect at playing a sport in their first game. Being an expert at a game takes practice, and more practice. So does doing science well. Best wishes, and please remember to mention us in your acceptance speech when you are awarded a Nobel Prize for your research.

Sincerely,

J.H.C.
R.N.G.
R.J.R.

Experimenting Safely

Wearing your seat belt in a car and using protective pads and a helmet when skateboarding make good sense. Similar safety precautions are also important when conducting a science project. **Before you begin your experiment, be sure your teacher has reviewed your procedures for safety.** If you are conducting your experiment at home, you should also discuss your safety precautions with your parents as well.

Safety concerns for different kinds of projects are described in separate sections of this appendix. These sections are: A) chemicals, B) mold and microorganisms, C) electricity, D) radiation, and E) animals and humans. Read the sections that are related to your project. **The safety guidelines here are only a sample. Be sure you understand and follow all the safety procedures needed for your own project.**

A. CHEMICALS

Cleaners, fertilizers, and other chemicals serve many useful purposes, but all of them can be dangerous if improperly used. Never mix chemicals, not even household cleaners, without help from an adult. In addition, you should:

- **Always wear protective glasses.** Gloves and an apron are also good ideas.
- **Wash your hands after handling any chemical.**
- **Know the potential dangers of the chemical you are using.** Some chemicals can irritate your skin, while others are poisonous.
  Do not breathe in vapors from chemicals.
  Be sure the area in which you are working is well-ventilated.
- **Know how and where to store chemicals safely.** A special kind of container might be needed, or maybe the chemical should be stored in a glass instead of a plastic container.
- **Know what to do in case of an accident.**
- **Know the procedures for safely disposing your chemicals.**

Your science teacher can help you find answers to safety questions in laboratory manuals or chemical catalogs, such as the *Flinn Chemical Catalog and Reference Manual.* Most schools also have information sheets on the chemicals used in science classes. These are called Material Safety Data Sheets, or MSDS.

B. MOLD AND MICROORGANISMS

You have probably seen mold growing on bread and other foods because molds are all around us. Microorganisms are also everywhere. Most common molds and microorganisms are harmless, but some are harmful. Before beginning any project with molds, ask your parents if you or anyone in your family is allergic to molds. Follow these safety precautions:

- **Keep the mold and microorganisms containers covered.**
- **Do not touch the molds or microorganisms.**
- **Wash your hands frequently.**
- **Never smell molds and microorganisms by inhaling close to the containers.**
- **Do not re-use containers.**
- **Dispose of your organisms and containers properly.**
- **Avoid growing molds on soil; some can make you sick.**

When growing molds and microorganisms, you will often grow "uninvited" molds, bacteria, fungi, and yeasts. Most of these are also harmless, but some are not. Play it safe. Properly dispose of these uninvited guests.

Similar care should be taken when studying other microorganisms such as bacteria, protozoa, and algae. Learn as much as you can about these organisms before beginning any experiment. Bacteria, for example, are often grown in special containers called petri dishes. Harmful bacteria as well as safe bacteria may grow in these containers. Follow the same safety procedures as those given for working with molds.

C. ELECTRICITY

Experiments that use electricity should always be checked by an adult who knows how to safely work with electricity. Take the proper precautions to prevent an accident. When designing and conducting your experiment, you should:

- **Use as little voltage as possible.**
- **Avoid using current from household outlets; use batteries instead.**
- **Watch for leaky batteries.** The chemicals inside can be harmful.
- **Make sure electrical appliances and tools are insulated and grounded.**
- **Never work alone.**

D. RADIATION

Experiments using microwave ovens, lasers, radon, and some types of smoke detectors all involve radiation—energy or streams of particles given off by atoms. Radiation can be very dangerous. Even in small amounts it can be harmful to living tissue.

Before beginning any experiment involving radiation, get help from someone who knows about the kind of radiation you would like to use in your experiment. Keep these safety precautions in mind:

- **Never work alone.**
- **Dispose of materials that give off radiation as required.**
- **Know the law. Certain state and federal laws may apply.**

E. ANIMALS AND HUMANS

If you plan to experiment with animals with backbones (vertebrates), you must follow very special rules. Vertebrates include fish, amphibians, reptiles, and birds as well as mammals. If you want to use vertebrates or their eggs, discuss your ideas with your teacher first. A qualified adult supervisor who is trained to take care of vertebrates, like a scientist or a teacher, must agree to begin supervising your project before you even obtain the first organism.

Most schools and competitions prefer that students use animals without backbones (*in*vertebrates) in animal experiments. Insects and worms are examples of invertebrates. If you do a project with animals, you must provide proper care for all the animals. Proper care includes:

- **a comfortable living place;**
- **procedures that do not injure the organism;**
- **enough food, water, warmth and rest;**
- **gentle handling;**
- **humane disposal or a proper home for organisms when your experiment is finished.**

If you are conducting an experiment that may be entered in a competitive event, such as a science fair, be sure you read and follow their rules on the use of animals in experiments. For a copy of the complete and *current* rules of the Intel International Science and Engineering Fair rules, see your science teacher, contact the following organization for a copy or download the rules from their website.

Intel International Science and Engineering Fair
Science Service Incorporated
1719 N St., N.W.
Washington, DC 20036
(202) 785-2255
http://www.sciser.org/weststs.htm

Special rules must also be followed in experiments using humans. Nothing may be done to humans that is likely to cause them harm. Participation should be voluntary. Some experiments, like those that just involve observing people, may not need special signed forms and procedures. Talk with your teacher about experiments involving humans. Scientists who wish to do experiments on humans or animals must have their research plans approved by a committee of fellow scientists. These rules are to help insure that human and animal subjects are treated properly.

SUMMARY

There are risks with everything we do. Taking proper precautions and using safe procedures can reduce these risks. Cooking, for example, can be dangerous. But you can cook safely by being careful and following safety procedures that reduce the danger. That's why people use potholders and keep pot handles pointed in toward the stove. When you conduct your science experiment, practice good safety procedures, too. Safety is no accident; plan for it.

CHAPTER 1

Developing Basic Concepts

Objectives

- Identify and define the basic concepts of experimental design.
  - hypothesis
  - dependent variable
  - control group
  - independent variable
  - constants
  - repeated trials
- Construct a simple diagram to communicate the major components of an experiment.
- Use a checklist to evaluate the basic parts of an experiment and to identify improvements.

National Standards Connections

- Identify questions that can be answered through scientific investigation (NSES).
- Systematically collect, organize, and describe data (NCTM).

What do designing a race car, creating a cake recipe, finding a flu vaccine, or simply answering the question "What is the best way to . . . ?" all have in common? One answer is that each of these tasks requires doing experiments, and lots of them. Learning how to design experiments that really work is a challenge, but you can do it. Experimenting requires knowledge and skills that most of us need but do not have. So let's begin learning how to design an experiment using a very simple system—a paper airplane.

DESIGNER PLANES

Begin by doing Investigation 1.1, *Designer Planes*.

TURN TO Investigation 1.1, page 2

In Investigation 1.1, *Designer Planes*, you and your classmates made airplanes from identical sheets of paper. Your class lined up to throw the planes to determine who had made the 'best' one. But, before you could throw your planes, your class had to select a single criterion for 'best' that would be used in judging the contest. The planes were thrown, and some flew better than others.

You were then directed to change your plane in some way. You were asked to predict how the change you purposely made in your airplane would affect its

INVESTIGATION 1.1 ▪ Designer Planes

Materials

- Paper
- Scissors
- Tape
- Paper clips
- Safety goggles

Safety

- Wear goggles.
- Do not throw planes at fellow students.
- Handle sharp objects safely.

Procedure

Part I

1. Make a paper airplane. Write your name on it.
2. Follow your teacher's directions for flying the plane.
3. Observe the plane's flight.

Part II

4. Use the materials to modify your plane.
5. Make a **hypothesis** about the effect of the change.
6. Follow your teacher's directions for flying the plane.
7. Observe the plane's flight.

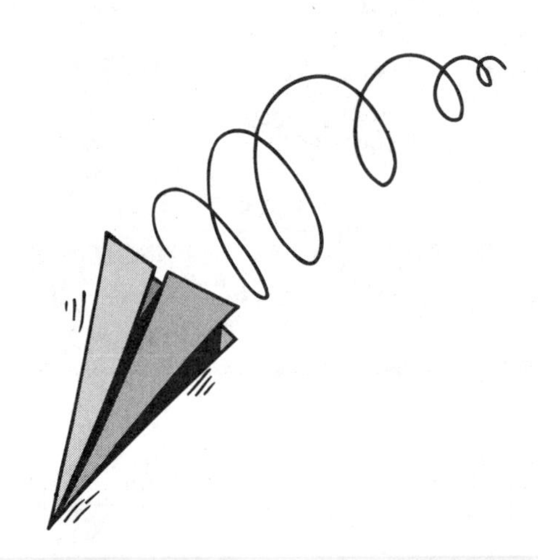

Thought Questions

1. How did you **act** upon your plane?
2. What did you **purposely change** about your plane?
3. How did you determine your plane's **response?**
4. What **remained the same** about your plane?

Class Data Table

Question 1	*Question 2*	*Question 3*	*Question 4*
Action	**Purposely Changed**	**Response to Change**	**Remained the Same**

flight using an **If . . . , then . . .** format. In an experiment a statement of your best thinking about how the change you purposely made might affect another factor is called a **hypothesis.** A hypothesis for the airplane investigation might be, "**If** a paper clip is added to the nose of an airplane, **then** it will fly straighter."

You lined up with your modified planes and were asked if a single throw, called a **trial,** was the fairest way of determining whose plane was best. You probably decided several throws would be better than a single throw. Scientists assume that each trial, however carefully done, contains a chance error or two. Some chance errors have no effect on the results. Other chance errors can result in an increase or a decrease in the result. They also assume that if enough trials are done, chance errors that increased the results will be balanced by the chance errors that decreased the results.

Having done Investigation 1.1, can you answer the ***Thought Questions***?

You and your classmates acted on your planes by adding paper clips, refolding, taping, and so on. Through those actions, you purposely changed something about the plane, such as its weight, center of balance, or wing design. In response to the change you made in your plane, it may have flown farther, straighter, or higher. Factors that remained the same were the size, texture, weight of the original paper, the flight line, and room conditions. The **Thought Questions** and the table at the bottom of Investigation 1.1 involve the concept of change. Synonyms for change are *modified, fixed, altered,* and *varied.* A long time ago, scientists selected the synonym *varied* and used the term **variable** to describe each factor that changes in an experiment. The variable that is purposely changed or manipulated in an experiment is called the **independent variable** and the variable that changes in response is called the **dependent variable.** Independent and dependent variables are often called the **manipulated variable** and the **responding variable.** The statement predicting the effect of the changes made in the independent variable on the dependent variable is called the **hypothesis.** For example, "**If** the plane is stapled in the middle, **then** it will fly farther." Similarly, a **title** for an experiment states, "**The Effect of** the Independent Variable **on** the Dependent Variable." For example, a title for an airplane experiment could be, "**The Effect of** Added Weight **on** the Distance an Airplane will Fly." Note that all the important words are capitalized in a title. All the factors that remain the same throughout the experiment are called **constants,** such as size, weight, texture of paper, flight line, direction of flight, and room conditions.

Unnoticed changes in the room's air flow from heating units, doors, or windows may have affected your plane's flight. How could you be sure that the change you purposely made in your plane really affected its flight, rather than some unknown factor? This is a very important concept. In doing an experiment there is no way to be absolutely sure that you are keeping every one of the constants from changing. Factors you thought were constants may actually be changing. You must establish a way to detect any hidden variables that may be changing without you knowing it. In the plane experiment, one way to detect hidden variables would be to fly at least one unmodified airplane in each trial to see whether its flight remained the same. If an unchanged plane flew the same in each flight, you could safely conclude that there were no hidden variables that affected the flight of your modified airplanes. The term **control** describes the unchanged plane that was used to detect and measure effects of hidden variables.

At this point, you should be able to define and identify the hypothesis, independent variable, dependent variable, constants, control, and repeated trials in an investigation. Without looking, try to define each term.

HOT SOLUTIONS

Nearly all solids dissolve better in hot water. But does the dissolving process affect the water temperature? Calcium chloride is a solid that dissolves easily in water. Calcium chloride is used to prevent icing of walkways and roads. Suppose you observed that a beaker of water warmed as the calcium chloride dissolved. What would you hypothesize about the effect of adding more calcium chloride to the water? In this case, a **hypothesis** would describe the effect of adding more calcium chloride on the change in temperature of the water. For example, "**If** more scoops of calcium chloride are added to water, **then** the temperature of the water will increase." Investigation 1.2, *Hot Solutions*, contains a list of materials and a procedure that you can use to test your hypothesis. Follow your teacher's directions. Be sure to use appropriate safety precautions including wearing safety goggles, washing your hands when you have finished, and disposing of the chemicals in properly marked containers. For more information about safety precautions when using chemicals, see Section A in ***Experimenting Safely*** in front matter.

In Investigation 1.2, *Hot Solutions,* identify the following:

- independent variable
- dependent variable
- levels of the independent variable
- constants
- control
- repeated trials
- hypothesis

Use the following format to write a title: "**The Effect of** the *Independent Variable* **on** the *Dependent Variable.*"

TURN TO Investigation 1.2, page 5

Sample data for *Hot Solutions* are provided in Table 1.1 *Class Data from Hot Solutions.* Each measurement is a separate trial. Remember that each trial may contain a chance error, increasing or decreasing its value. Do you have more confidence in one measurement or the average of several measurements? Why? Averages of repeated trials increase confidence by reducing the effects of chance or random errors that often occur in a single trial. What varied or changed in the experiment? Remember, the term **variable** describes all factors that change or could change in an experiment. The variable that you purposely changed or manipulated, the amount of chemical, is called the **independent or manipulated variable.** The values of the independent variable that you test are called the **levels of the independent variable.** In this case, the levels of the independent variable are 0, 1, 2, and 3 scoops. The variable that responded, the temperature of the water, is called the **dependent or responding variable.** What factors remained the same in the experiment? Amount of water (75 ml), time to dissolve (2 minutes), and stirring are examples; label all the factors that remained the same as **constants.**

TABLE 1.1 Class Data from Hot Solutions

Amount of chemical (scoops)	Change in temperature (°C)			Average change in temperature (°C)
	Trials			
	1	2	3	
0	0	1	1	1
1	8	9	7	8
2	15	11	14	13
3	20	17	17	18

INVESTIGATION 1.2 ▪ Hot Solutions

Materials

- 150 ml beaker or plastic cup
- Thermometer (°C)
- Scoop or plastic spoon
- Safety goggles
- Chemical (calcium chloride)
- Clock
- Water
- Graduated cylinder

Safety

- Wear goggles.
- Wash hands.
- Dispose of chemicals in marked containers.

Record & Calculate

°C End Temp
− °C Initial Temp
°C Change in Temp

°C End Temp
− °C Initial Temp
°C Change in Temp

°C End Temp
− °C Initial Temp
°C Change in Temp

°C End Temp
− °C Initial Temp
°C Change in Temp

°C End Temp
− °C Initial Temp
°C Change in Temp

°C End Temp
− °C Initial Temp
°C Change in Temp

°C End Temp
− °C Initial Temp
°C Change in Temp

°C End Temp
− °C Initial Temp
°C Change in Temp

°C End Temp
− °C Initial Temp
°C Change in Temp

°C End Temp
− °C Initial Temp
°C Change in Temp

°C End Temp
− °C Initial Temp
°C Change in Temp

°C End Temp
− °C Initial Temp
°C Change in Temp

Procedure

1. Place 75 ml (about 1/3 cup) of water in a beaker or cup.
2. Record the initial temperature of the water (°C) *in the margin.*
3. Measure the designated amount of the calcium chloride, such as 0, 1, 2, 3 scoops or spoonfuls.
4. Add and stir the designated amount of the chemical for 2 minutes.
5. Record the temperature (°C) of the water at the end of 2 minutes *in the margin.*
6. Calculate the temperature change of the water.
7. Record data on the class data table.

Amount of chemical (scoops)	Change in temperature (°C)			Average change in temperature (°C)
	Trials			
	1	2	3	
0				
1				
2				
3				

How could you improve this experiment?

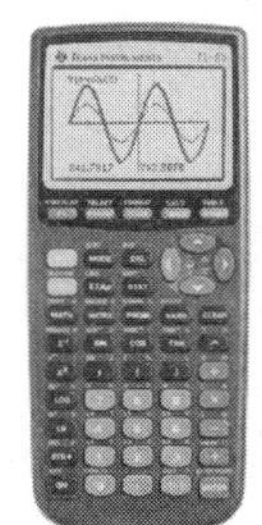

USING TECHNOLOGY

1. In the **STAT** mode of your calculator, enter the number of scoops in List 1 and the values for mean change in temperature in List 2. (See Appendix A, *Using Technology*, for additional help in using the graphing calculator).
2. In setting up your graph, select scatter plot as your graph type and List 1 for your x values and List 2 for your y values. Graph the data.
3. Press **Trace** and use the arrow keys to highlight each x value (number of scoops) and to see the corresponding y values (temperature).

Why were measurements made on water containing 0 scoops of calcium chloride? In this experiment, the cups receiving no chemical were the control and were used to detect or measure the influence of hidden variables, such as possible temperature gain from the room or from stirring. In some experiments the control is the group that receives no treatment, such as the cups receiving 0 scoops of calcium chloride. In other experiments, one of the levels of the independent variable may be selected as the control, usually the normal or typical case. For example, in experiments on detergents the *recommended amount* could be used as the control. In an experiment on the effect of 5, 10, 15, and 20 ml of water on plant growth, 0 ml of water would make no sense. The plants would die if a 'no treatment' control was used. However, if 10 ml of water was selected as the control because it was the normal or typical amount of water given to plants, then 10 ml would be the control. Any large variation in the growth of plants receiving 10 ml of water might suggest the influence of a hidden variable. In still other experiments, the control may be a standard outside the samples being tested. In testing rugs, wool rugs are universally used as the standard.

Related Web Sites

http://pointer.wphs.K12.va.us/118sci.htm (teacher adaptation/use components of an experiment)

http://www.isd77.k12.mn.us/resources/cf/SciProjInter.html (discussion of repeated trials, random errors, and systematic errors)

Practice

In the "Floor Wax Test" scenario below, identify the following components of an experiment:

1. independent variable
2. dependent variable
3. constants
4. repeated trials
5. control

Also using the scenario below write a title and a hypothesis using the following formats:

6. Title: The Effect of the *(changes in the independent variable)* on the *(dependent variable).*

7. Hypothesis: If the *(independent variable—describe how it will be changed),* then the *(dependent variable—describe the effect).*

A shopping mall wanted to determine whether the more expensive "Tough Stuff" floor wax was better than the cheaper "Steel Seal" floor wax at protecting its floor tiles against scratches. One liter of each grade of floor wax was applied to each of 5 test sections of the main hall of the mall. The test sections were all the same size and were covered with the same kind of tiles. Five (5) other test sections received no wax. After 3 weeks, the number of scratches in each of the test sections was counted.

CHAPTER 2

Applying Basic Concepts

Objectives

- Identify the basic concepts of experimental design in a scenario of an experiment.

hypothesis	independent variable
dependent variable	constants
control group	repeated trials

- Draw an experimental design diagram for an experimental scenario.
- Use a checklist to evaluate the design of an experiment.
- Suggest ways to improve the design of an experiment.

National Standards Connection

- Identify questions and concepts that guide scientific investigation (NSES).

Lists can be useful tools of analysis, but diagrams are frequently much more powerful tools. You could analyze an experiment, for example, by listing its components. However, constructing an experimental design diagram of the same experiment that states its title, hypothesis, independent variable, dependent variable, constants, control, and number of repeated trials is a more effective way to quickly visualize the design of an experiment. Figure 2.1 *Experimental Design Diagram* summarizes Investigation 1.2, *Hot Solutions* from Chapter 1.

To construct an experimental design diagram for any experiment with one independent variable, follow these steps:

1. Write a **title: The Effect of** the Independent Variable **on** the Dependent Variable.
2. State a **hypothesis: If** the (independent variable) is (describe how you changed it), **then** the (dependent variable) will (describe the effect).
3. Write your **independent variable (IV:).**
4. Divide the bottom two rows into columns; one column for each **level** of the independent variable. Write the levels of the IV in the columns. If one of those levels is the control for the experiment, put the word **(control)** under that level.

Title: The Effect of Various Amounts of Calcium Chloride on the Temperature of Water
Hypothesis: If more scoops of calcium chloride are added to water, then the temperature of the water will increase.

IV: Amount of calcium chloride (scoops)				
0 scoop (control)	1 scoop	2 scoops	3 scoops	← Levels of Independent Variable Including the Control
3 trials	3 trials	3 trials	3 trials	← Repeated Trials Number of times each of the levels of IV was tested

← Independent Variable (IV row)

DV: Temperature of water ← Dependent Variable

C: Same amount of water (75 ml) ← Constants
Same time to dissolve (2 min)
Constant stirring

Figure 2.1 Experimental Design Diagram.

5. In each column write the number of **repeated trials** conducted for each level of the independent variable.

6. Put your **dependent variable (DV:)** below the rectangle.
7. Write a list of the **constants (C:)**.

CHECKING THE DESIGN

Because each part of an experiment has its place in an experimental design diagram, you can spot missing or weak parts quickly and easily. Using the checklist in Table 2.1 *Checking the Experimental Design* begin at the top of the experimental design diagram and look for ways to improve the experiment.

Using the checklist questions to critique the experimental design diagram for the *Hot Solutions* experiment (Figure 2.1) produced the following results:

Q1. Does the title clearly identify both the independent and dependent variables?
The original title, Hot Solutions, *does not identify the independent and dependent variables. This title does,* The Effect of Various Amounts of Calcium Chloride on the Temperature of Water.

Q2. Does the hypothesis clearly state how you think changing the independent variable will affect the dependent variable?
The hypothesis is fine.

Q3. Is there an independent variable?
The amount of calcium chloride is a single, well-defined independent variable.

Q4. Are the levels of the independent variable clearly stated? Are there enough levels of the independent variable tested?
Yes, the levels are sufficient and clearly stated.

Q5. Is there a control? Is it clearly stated?
There is a control, 0 scoops of calcium chloride.

Q6. Are there repeated trials? Are there enough of them?
Three trials are ok, but 5 or more would have been better. A common question is, "How many trials are enough?" The answer depends on the experiment. There are few differences among granules of calcium chloride or among paper airplanes that are constructed the same way. When you inves-

TABLE 2.1 Checking the Experimental Design

Checklist Question Number	Questions
☐ 1.	Does the title clearly identify both the independent and dependent variables?
☐ 2.	Does the hypothesis clearly state how you think changing the independent variable will affect the dependent variable?
☐ 3.	Is there just one independent variable? Is it well defined?
☐ 4.	Are the levels of the independent variable clearly stated? Are there enough levels of the independent variable tested? Are there too many?
☐ 5.	Is there a control? Is it clearly stated?
☐ 6.	Are there repeated trials? Are there enough of them?
☐ 7.	Is the dependent variable clearly identified and stated?
☐ 8.	Is the dependent variable operationally defined? Operationally defined means that the investigator very clearly stated how the response would be measured or described?
☐ 9.	Are the constants clearly identified and described? Are there any other?
☐ 10.	Did the experimental design diagram include all the parts listed in Figure 2.1? Were all the parts placed in the proper place?
☐ 11.	Was the experiment creative? Was it an appropriate level of complexity?

tigate such non-living things, you tend to get very similar results. For this reason, you can use a smaller number of trials, such as 5 to 10. However, there are many differences among even similar looking groups of plants and animals. Because of the many differences in organisms, you tend to get a greater variety of results; therefore, more trials are needed. With human studies, such as studies on taste preferences, even more trials are necessary. With organisms you should generally conduct as many trials as time, money, and space will allow.

Q7. Is the dependent variable clearly identified and stated?
The temperature of the water is not the dependent variable. The change in the temperature of the water is.

Q8. Is the dependent variable operationally defined? Operationally defined means that the investigator very clearly stated how the response would be measured or described.
No, the experimental design diagram did not indicate that the water temperature was measured in degrees Celsius. To improve the diagram, add the symbol (0 °C) to show how the change in water temperature was measured.

For the dependent variable, you can use either quantitative or qualitative data. ***Quantitative*** *observations include measurements with standard scales, such as degrees Celsius.* ***Qualitative*** *observations include verbal descriptions or measurements with non-standard scales. For example, you could describe the cloudiness or color of the solution. You will learn more about these types of dependent variables in Chapter 8,* Analyzing Experimental Data.

Q9. Are the constants clearly identified and described? Are there any others?
Some constants are clearly identified, such as the amount of water (75 ml) and the time to dissolve (2 min). Other constants are unclear. For example, what is the scoop size? Is it a level or heaping scoop? It would be better to use a balance to measure the calcium chloride in grams. What about the initial

temperature of the water for each trial? Was it always the same? If not, this could be a hidden variable affecting the results. What is meant by constant stirring? How could you be sure the stirring was the same?

Other unlisted factors may also have affected the results of the experiment. For example, what type of container was used? Were the containers good or poor insulators? How might the type of container affect the results? What other factors can you identify that might have affected the experimental results?

Q10. Did the experimental design diagram include all the parts listed in Figure 2.1? Were all the parts placed in the proper place?
Yes, all the parts were included and properly placed.

Q11. Was the experiment creative? Was it at an appropriate level of complexity?
Deciding upon creativity and appropriateness are 'judgement calls' on your part. For an experiment to be appropriate, it should address a question whose answer is unknown to the investigator. Determining the effect of different colors of ground covers on plant growth is an appropriate experiment, while determining the effect of light versus dark on plant growth is not. Almost everyone knows that plants will die without light. In determining appropriateness, you will also need to evaluate your own background knowledge. What would have been an appropriate experiment for you in sixth grade is not an appropriate experiment for your junior year.

Use the checklist in Table 2.1 to evaluate the experimental design diagrams you construct for your experiments. Ask a friend or family member to check your diagram as well. Use their comments to improve your design and then submit it to your teacher for review.

At this point you should be able to describe the major parts of an experiment. You should also be able to identify the basic parts of an experiment in the investigation that you conducted on dissolving various amounts of calcium chloride in water. Perhaps, you can even draw a good experimental design diagram that would score high on a checklist, but what about designing your own experiment? Are you ready?

You may need some practice in identifying and diagramming the basic parts of an experiment. You also need to be able to identify common errors, such as no control group, poorly defined constants, and too few trials. There are also unique experimental features, such as proper use of animals, that you should learn more about.

One way to practice these new ideas is to read examples of experiments. These examples can be investigations in your textbook, experiments conducted by other students, or brief descriptions of experiments called scenarios. In this chapter you will use scenarios of experiments to refine your skills.

LEARNING FROM SCENARIOS

Five scenarios of experiments are included in Activity 2.1, *Design Detective*. For each scenario, your first task is to read the description of the experiment and to identify the major parts, such as the independent variable and the number of repeated trials. Then, complete the following:

1. Draw an experimental design diagram for the experiment described in the scenario. Include only the parts specifically described in the scenario. If you need assistance, refer to Table 1.2, *General Format for an Experimental Design Diagram*.
2. Evaluate the experimental design diagram using the checklist in Table 2.1.
3. Make a list of ways you could improve the experiment described in the scenario. Modify the experimental design diagram to include the improvements.

Finally, check your work with the experimental design diagram and discussion provided for each scenario. Complete one scenario at a time. With each scenario, you will learn new ways to improve the design of an experiment. Resist the urge to read ahead before you do your part; do not cheat yourself out of understanding good experimental design.

When you do compare your suggestions to those provided, yours may be different. Do not worry because there are many ways to improve the design of these investigations. Your task is to suggest as many ways as you can. Share your suggestions with classmates and your teacher; they may have different suggestions for improvement.

TURN TO Activity 2.1, page 14

COMPOST AND BEAN PLANTS

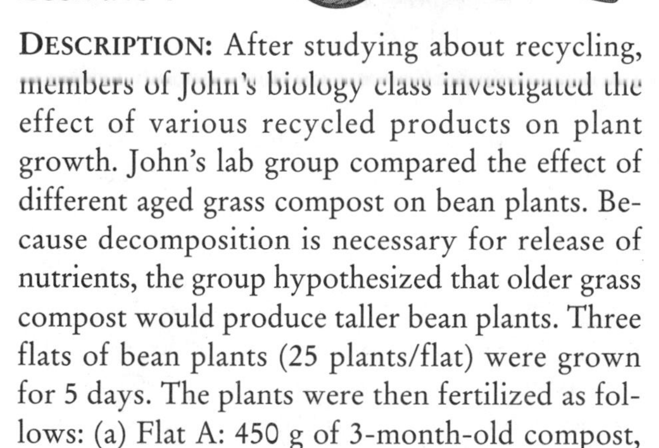

Scenario 1

DESCRIPTION: After studying about recycling, members of John's biology class investigated the effect of various recycled products on plant growth. John's lab group compared the effect of different aged grass compost on bean plants. Because decomposition is necessary for release of nutrients, the group hypothesized that older grass compost would produce taller bean plants. Three flats of bean plants (25 plants/flat) were grown for 5 days. The plants were then fertilized as follows: (a) Flat A: 450 g of 3-month-old compost, (b) Flat B: 450 g of 6-month-old compost, and (c) Flat C: 0 g compost. The plants received the same amount of sunlight and water each day. At the end of 30 days the group recorded the height of the plants (cm).

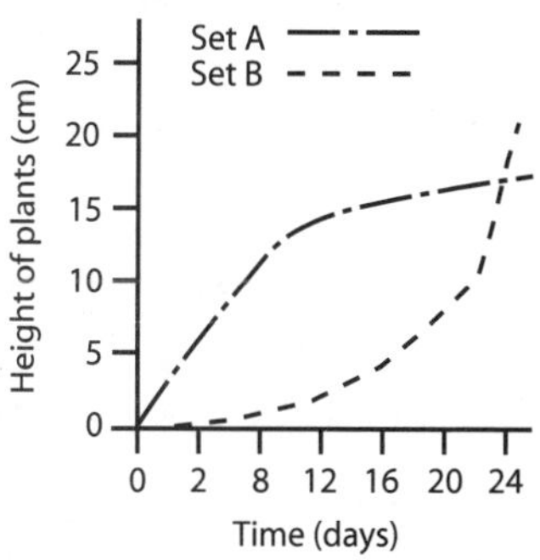

DISCUSSION: In Scenario 1, did you infer that plant species and type of soil are constants? The description neglects to give this information. It is important to clearly state all the constants in an experiment. Can you think of ways compost might affect plants other than their height? Compost affects plant growth in many ways including color of leaves, number of flowers or fruit, size of leaves, and sturdiness of stems. In Chapter 8 you will learn more about both quantitative and qualitative measurements. Many first time researchers make too few measurements over too short a time period and that can be a problem. Consider the graph of data on two groups of plants with the same average height at 24 days, but different average heights at 12 days. Can you interpret the data and make recommendations for fertilizer use? How many measurements of plant height would you make to feel confident about your recommendation?

Title: The Effect of Different Aged Compost on Bean Plant Growth
Hypothesis: If older compost is applied, then plant growth will be increased.

IV: Age of Compost		
3-month-old compost	6-month-old compost	No compost (Control)
25 Plants	25 Plants	25 Plants

DV: Height of plants (cm)

C: Amount of light
Amount of water
Amount of compost

Figure 2.2 Experimental Design for Plants and Compost.

ACTIVITY 2.1 ▪ Design Detective

SCENARIO 1 ▪ Compost and Bean Plants

After studying about recycling, members of John's biology class investigated the effect of various recycled products on plant growth. John's lab group compared the effect of different aged grass compost on bean plants. Because decomposition is necessary for release of nutrients, the group hypothesized that older grass compost would produce taller bean plants. Three flats of bean plants (25 plants/flat) were grown for 5 days. The plants were then fertilized as follows: (a) Flat A: 450 g of 3-month-old compost, (b) Flat B: 450 g of 6-month-old compost, and (c) Flat C: 0 g compost. The plants received the same amount of sunlight and water each day. At the end of 30 days the group recorded the height of the plants (cm).

SCENARIO 2 ▪ Metals and Rusting Iron

In chemistry class, Allen determined the effectiveness of various metals in releasing hydrogen gas from hydrochloric acid. Several weeks later, Allen read that a utilities company was burying lead next to iron pipes to prevent rusting. Allen hypothesized that less rusting would occur with the more active metals. He placed the following into four separate beakers of water: (a) 1 iron nail, (b) 1 iron nail wrapped with an aluminum strip, (c) 1 iron nail wrapped with a magnesium strip, (d) 1 iron nail wrapped with a lead strip. He used the same amount of water, equal amounts (mass) of the metals, and the same type of iron nails. At the end of 5 days, he rated the amount of rusting as small, moderate, or large. He also recorded the color of the water.

SCENARIO 3 ▪ Perfumes and Bees' Behavior

JoAnna read that certain perfume esters would agitate bees. Because perfume formulas are secret, she decided whether to determine whether the unknown Ester X was present in four different perfumes by observing the bees' behavior. She placed a saucer containing 10 ml of the first perfume 3 m from the hive. She recorded the time required for the bees to emerge and made observations on their behavior. After a 30-minute recovery period, she tested the second, third, and fourth perfumes. All experiments were conducted on the same day when the weather conditions were similar; that is, air, temperature, and wind.

SCENARIO 4 ▪ Fossils and Cliff Depth

Susan observed that different kinds and amounts of fossils were present in a cliff behind her house. She wondered if changes in fossil content occurred from the top to the bottom of the bank. She marked the bank at five positions: 5, 10, 15, 20, and 25 m from the surface. She removed 1 bucket of soil from each of the positions and determined the kind and number of fossils in each sample.

SCENARIO 5 ▪ *Aloe vera* and Planaria

Jackie read that *Aloe vera* promoted healing of burned tissue. She decided to investigate the effect of varying amounts of *Aloe vera* on the regeneration of planaria. She bisected the planaria to obtain 10 parts (5 heads and 5 tails) for each experimental group. She applied concentrations of 0%, 10%, 20%, and 30% *Aloe vera* to the groups. Fifteen milliliters of *Aloe vera* solutions were applied. All planaria were maintained in a growth chamber with identical food, temperature, and humidity. On Day 15, Jackie observed the regeneration of the planaria parts and categorized development as full, partial, or none.

METALS AND RUSTING IRON

Scenario 2

DESCRIPTION: In chemistry class, Allen determined the effectiveness of various metals in releasing hydrogen gas from hydrochloric acid. Several weeks later, Allen read that a utilities company was burying lead next to iron pipes to prevent rusting. Allen hypothesized that less rusting would occur with the more active metals. He placed the following into four separate beakers of water: (a) 1 iron nail, (b) 1 iron nail wrapped with an aluminum strip, (c) 1 iron nail wrapped with a magnesium strip, (d) 1 iron nail wrapped with a lead strip. He used the same amount of water, equal amounts (mass) of the metals, and the same type of iron nails. At the end of 5 days, he rated the amount of rusting as small, moderate, or large. He also recorded the color of the water.

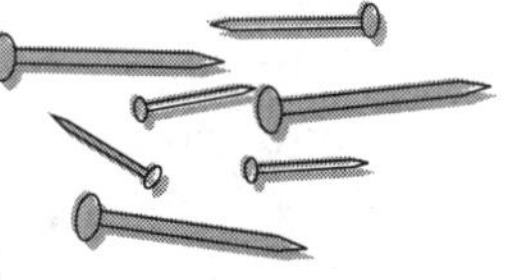

DISCUSSION: An excellent design feature is the presence of a control group, the iron nail without a metallic strip. The number of trials, 1 nail, is insufficient and should be increased to five or more. The number of trials required for sufficiency is determined by the variability in the experimental organisms or materials. With physical phenomena, fewer trials are necessary than for living organisms. The dependent variable, amount of rusting, should be quantified. Possibilities include measuring the residue obtained by scraping the nails or measuring the mass of the nails before and after the experiment. Both the colors of the water and the residue (precipitate) should be recorded. Color observations are critical clues to chemical reactions. In this case, they indicate the kind of iron products formed. Keeping the mass of the metallic wrapper constant may have introduced an undetected variable, the surface area of the iron nail exposed to the water and oxygen, into the experiment. Because chemical reactions occur at the interfaces of substances, keeping surface area constant and varying the mass would be a critical way to improve the experimental design. The type of water, distilled, tap, and so on, should also be described. For information about safety precautions when using chemicals, see Section A *Experimenting Safely* in front matter.

Title: The Effectiveness of Various Metals in Preventing the Rusting of Iron
Hypothesis: If the chemical activity of the metallic wrapper is increased, then less rusting of iron will occur.

IV: Type of metallic wrapping strip			
Iron nail with no metal (Control)	Iron nail with magnesium	Iron nail with aluminum	Iron nail with lead
1 Trial	1 Trial	1 Trial	1 Trial

DV: Amount of rusting
Color of water

C: Amount of water
Mass of metallic wrapper
Type of iron nail

Figure 2.3 Experimental Design Diagram for Metals and Rusting Iron.

PERFUMES AND BEES' BEHAVIOR

Scenario 3

DESCRIPTION: JoAnna read that certain perfume esters would agitate bees. Because perfume formulas are secret, she decided to determine whether the unknown Ester X was present in four different perfumes by observing the bees' behavior. She placed a saucer containing 10 ml of the first perfume 3 m from the hive. She recorded the time required for the bees to emerge and made observations on their behavior. After a 30-minute recovery period, she tested the second, third, and fourth perfumes. All experiments were conducted on the same day when the weather conditions were similar, that is, air, temperature and wind.

DISCUSSION: Significant ways to improve JoAnna's experiment are to increase the number of trials and to add a control group such as a nonfragrant liquid (water). When animal subjects are used, be sure to allow sufficient recovery time between trials. Because one perfume might influence reaction to another perfume, presentation order should be randomized. With four bee hives, a sample presentation order might be

Trial	Perfume
Trial 1	1 ..... 2 ..... 3 4
Trial 2	2 ..... 3 ..... 4 1
Trial 3	3 ..... 4 ..... 1 2
Trial 4	4 ..... 3 ..... 2 1

You should be able to justify the constants in the experiment. "Why should the temperature be kept the same?" "Why are wind conditions important?" You should also be able to describe the animal's normal behavior. For example, baseline data could be collected by observing the bees' frequency of emergence and behavior for several days prior to exposure to the liquids.

Do you find it strange that the 25 plants in Scenario 1 constitute repeated trials; whereas, the hundreds of bees in Scenario 3 constitute a single trial? The reason is that each plant is measured independently, giving 25 individual points of data for compost of each age; whereas, each hive is a single unit that yields only one piece of data, the time required for the bees to emerge. In Scenario 1 there are 25 points of data to average. In Scenario 3 one cannot compute an average because only one trial was conducted. For information about safety precautions when using animals, see Section E in ***Experimenting Safely*** in front matter.

Title: The Effect of Various Perfumes on the Behavior of Bees

Hypothesis: If the perfume contains Ester X, then the bees will display agitated behavior.

IV: Type of perfume			
Perfume 1	Perfume 2	Perfume 3	Perfume 4
1 Trial	1 Trial	1 Trial	1 Trial

DV: Time to emerge
Behavior of bees

C: Amount of perfume (10 ml)
Distance from hive (3 m)
Weather Conditions (air, temperature, wind)

Figure 2.4 Experimental Design Diagram for Bees and Perfume.

FOSSILS AND CLIFF DEPTH

Scenario 4

DESCRIPTION: Susan observed that different kinds and amounts of fossils were present in a cliff behind her house. She wondered if changes in fossil content occurred from the top to the bottom of the bank. She marked the bank at five positions: 5, 10, 15, 20, and 25 m from the surface. She removed 1 bucket of soil from each of the positions and determined the kind and number of fossils in each sample.

DISCUSSION: The number of trials, 1 bucket, is insufficient. Five or more samples could be drawn randomly from across the bank at the same depth. Recording both the numbers and kinds of fossils as dependent variables is an excellent feature that broadens the options for data analysis. Susan's experiment is a good example of how a nonexperimental project could be changed into an experimental project. By treating the depth from the surface as an independent variable, an **ex post facto experiment** or experiment after the fact was created. No control group exists in the experiment unless arbitrarily defined by the researcher. For example, if the literature review indicated that different types of fossils might occur above and below the 15 m level, it could be defined as the control. These designs are especially appropriate in earth science where relationships and patterns in events that occurred long ago are studied. This is also true for psychology where human characteristics, such as gender, intelligence, and age, are used to divide data into subgroups.

ALOE VERA AND PLANARIA

Scenario 5

DESCRIPTION: Jackie read that *Aloe vera* promoted healing of burned tissue. She decided to investigate the effect of varying amounts of *Aloe vera* on the regeneration of planaria. She bisected the planaria to obtain 10 parts (5 heads and 5 tails) for each experimental group. She applied concentrations of 0%, 10%, 20%, and 30% *Aloe vera* to the groups. Fifteen ml of *Aloe vera* solutions were applied. All planaria were maintained in a growth chamber with identical food, temperature, and humidity. On Day 15, Jackie observed the regeneration of the planaria parts and categorized development as full, partial, or none.

Title: The Effect of Bank Position on Fossils

Hypothesis: As you move from the top to the bottom of the cliff, then fossil content will decrease.

IV: Depth of sample from surface				
5 m	10 m	15 m	20 m	25 m
1 Trial	1 Trial	1 Trial	1 Trial	1 Trial

DV: Kinds of fossils
Number of fossils

C: Same amount of soil (1 bucket)

Figure 2.5 Experimental Design Diagram for Fossils and Cliff Depth.

Title: The Effect of Various Concentrations of *Aloe vera* on the Regeneration of Planaria
Hypothesis: Higher concentrations of *Aloe vera* will increase the regeneration of planaria.

IV: Concentration of *Aloe vera*			
0%	10%	20%	30%
10 Trials	10 Trials	10 Trials	10 Trials

DV: Regeneration of parts (heads and tails)
C: Amount of solution (15 ml)
Growth conditions (temperature, humidity, food)

Figure 2.6 Experimental Design Diagram for *Aloe vera* and Planaria.

DISCUSSION: The independent variable for the experiment is the concentration of *Aloe vera.* The dependent variable is the regeneration of the planaria. A control group that received no *Aloe vera* was included; the amount of solution and growth conditions were kept the same. With living organisms, a preliminary experiment and library research should be conducted to determine the deadly amount that should not be exceeded. Follow guidelines for humane treatment of organisms published by professional organizations. ***Experimentation on vertebrates is discouraged and should be prohibited except when supervised by a mentor.*** For information about safety precautions when using animals, see Section E in *Experimenting Safely* in front matter.

EVALUATING YOUR EXPERIMENTAL DESIGN SKILLS

Use Table 2.2, *Checklist for Evaluating an Experimental Design Diagram*, to help you identify major flaws or errors in an experiment.

PRACTICING YOUR SKILLS

After you have completed Activity 2.1, *Design Detective*, apply your skills to some of the following practice problems. What flaws can you find? How could you improve the experimental design? Which practice problems appeal to you as potential experiments to conduct?

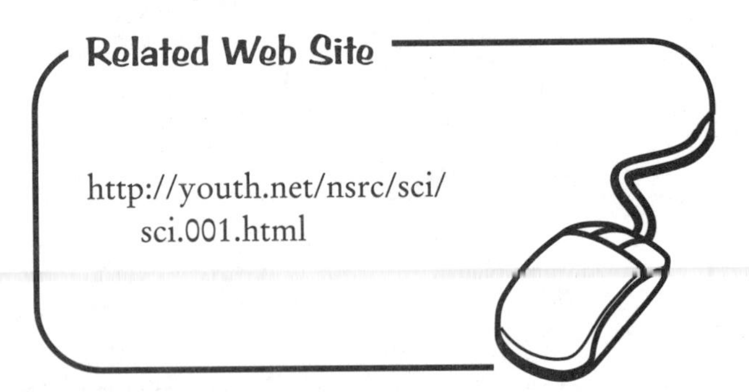

TABLE 2.2 Checklist for Evaluating an Experimental Design Diagram

Criteria	Self	Peer/Family	Teacher
Title			
Hypothesis			
Independent variable			
Levels of independent variable			
Control			
Repeated trials			
Dependent variable			
Operational definition of dependent variable			
Constants			
Creativity/Complexity			

Practice

For each of the scenarios below answer questions A–D.

A. Identify the independent variable, levels of the independent variable, dependent variable, number of repeated trials, constants, and control (if present).
B. Identify the hypothesis for the experiment. If the hypothesis is not explicitly stated, write one for the scenario.
C. Draw an experimental design diagram, which includes an appropriate title and hypothesis.
D. State at least two ways to improve the experiment described in the scenario.

1. Ten seeds were planted in each of 5 pots found around the house that contained 500 g of "Pete's Potting Soil." The pots were given the following amounts of distilled water each day for 40 days: Pot 1, 50 ml; Pot 2, 100 ml; Pot 3, 150 ml; Pot 4, 200 ml; Pot 5, 250 ml. Because Pot 3 received the recommended amount of water, it was used as a control. The height of each plant was measured at the end of the experiment.
2. Gloria wanted to find out if the color of food would affect whether kindergarten children would select it for lunch. She put food coloring into 4 identical bowls of mashed potatoes. The colors were red, green, yellow and blue. Each child chose a scoop of potatoes of the color of their choice. Gloria did this experiment using 100 students. She recorded the number of students that chose each color.
3. Susie wondered if the height of a hole punched in the side of a quart-size milk carton would affect how far from the container a liquid would spurt when the carton was full of the liquid. She used 4 identical cartons and punched the same size hole in each. The hole was placed at a different height on one side of each of the containers. The height of the holes were 5, 10, 15 and 20 cm from the base of the carton. She put her finger over the hole and filled the carton to a height of 25 cm with a liquid. When each carton was filled, she placed it in the sink and removed her finger. Susie measured how far away from the carton's base the liquid had squirted when it hit the bottom of the sink.
4. Sandy heard that plants compete for space. She decided to test this idea. She bought a mixture of flower seeds and some potting soil. Into each of 5 plastic cups she put the same amount of soil. In the first cup she planted 2 seeds, in the second cup she planted 4 seeds, in the third cup 8 seeds, and in the fourth cup she planted 16 seeds. In the last cup she planted 32 seeds. After 25 days, she determined which set of plants looked best.
5. Esther became interested in insulation while her parent's new house was being built. She decided to determine which insulation transferred the least heat. She filled each of 5 jars half-full with water. She sealed each jar with a plastic lid. Then she wrapped each jar with a different kind of insulation. She put the jars outside in the direct sunlight. Later, she measured the temperature of the water in each jar.

CHAPTER 3

Generating Experimental Ideas

Objectives

- Use the Four Question Strategy and a prop to brainstorm numerous variables, constants, and hypotheses for experiments.
- Describe a variety of props for use in brainstorming: general topics, lists of materials, science articles, questions, demonstrations, and text-book laboratory activities.
- Use a checklist to evaluate your responses to the Four Question Strategy.

National Standards Connections

- Identify questions that can be answered through scientific investigation (NSES).
- Design and conduct scientific investigations (NSES).

The fear of the blank page frequently paralyzes the thought processes of would-be-poets when they are suddenly asked to write a poem. This same fear may grip you when you are asked to generate an original experiment for a class assignment or for a science competition. It does not matter how well you have memorized the scientific method or even the definitions of variables, controls, and repeated trials. The result is always the same—panic and random thoughts. "What problem should I investigate? Why can't I think of a problem? Hey! I've got it! But, who cares? What about . . .? No, that would take too long and besides I don't have. . . ." And so it goes.

Most successful writers have learned long ago to talk through a topic, list ideas, make brief notes, write a draft, then revise and edit. These writers know that good writing does not begin by trying to write the final product. First, a rough draft is made to get the major points on paper. Spit and polish comes later.

Similarly, you should explore the possible variations of a research topic before you attempt to state a problem, a hypothesis, variables, constants, and the control. In this chapter, you will learn to brainstorm ideas for an original experiment.

Students assigned the task of bringing a specific research problem to class often propose broad topics such as plants or electricity. They frequently do not understand what the term *specific research problem* means to a scientist. Even experienced graduate students in college find the task of determining a specific research problem very difficult. Their first drafts are often too general and must be rewritten.

In the past, some students were given sources of well-defined projects that they often followed as a recipe. Later, when a judge asked them why a certain method was used, they could only reply, "That's what the book said to do." You and other students do not need recipes. You need a strategy to help you develop an interesting topic into a well-designed experiment. You also need that strategy modeled and practiced several times before you design an original experiment of your own.

THE FOUR QUESTION STRATEGY

Realistically, how can you change a general topic into a quality original research project? Because the typical idea of a specific research problem is a general topic like plants, it will be used to introduce the strategy to you. Begin by reading the following sequence of four questions for generating experiment ideas from a general topic such as plants.

Q1: What materials are readily available for conducting experiments on (plants) ?
Soils
Plants
Fertilizers
Water
Light/heat
Containers

You might have also listed warmth and other environmental conditions that plants need. The more things you list in your responses to Question 1, the better an experiment you will be able to design. Choose materials that are inexpensive and easy to find. You may be able to borrow materials from your school, parents, or people in the community. Next, ask yourself Question 2.

Q2: How do (plants) act?
Plants grow

The major thing that plants do is grow, but you may have brainstormed other actions as well. Plants also flower, wilt, produce fruit, and die. The action you choose will determine what your dependent variable will be when you get to Question 4. Question 4 is about how you will measure the action you choose. Continue on to Question 3.

Q3: How can I change the set of (plant) materials to affect the action?

Water
Amount
Scheduling
Method of application
Source
Composition
pH

Plants
Kind
Spacing
Age
Size

Containers
Location of holes
Number of holes
Shape
Material
Size
Color

Responses to Question 3 are possible variables you could choose when designing an experiment. The longer the lists, the more choices you will have. For water, maybe you also thought of temperature and time of watering. Did you think of any other ways to change plants or containers? Each variable you generate is a potential independent variable. When you choose a variable to manipulate, for example, amount of water, it becomes your independent variable. All the rest of the potential independent variables must become constants in your experiment. Assign each of these constants a value and take care that they do not vary in your experiment.

In addition to the responses listed above you might also have added these:

Soil	***Seeds***
Composition	*Size*
Amount	*Color*
Depth	*Number*
Compaction	*Planting depth*
	Age

Similarly, you could also develop lists for light, fertilizer, and environmental conditions. The last question in the *Four Question Strategy* is next.

Q4: How can I measure or describe the response of __(plants)__ to the change?
Count the number of leaves
Measure the length of the longest stem
Count the number of flowers
Determine the rate of growth
Mass (weight) of the fruit produced
Measure the diameter of the stems

This final question helps you decide how to measure or describe changes in the dependent variable you selected from Question 2. You might have also added: measure root development, record color, assess health quality, or still other ways to measure plant growth.

To design an experiment for a science project, all you have to do is select an **independent variable** from your responses to Question 3, such as *amount of water.* Then select a **dependent variable** from Question 4, such as *number of flowers.* To make your experiment a fair test of the effect of amount of water on the number of flowers produced, all other responses to Question 3 must be kept the same. They become **constants** if you assign a value to each and keep their value the same throughout your experiment.

You can write a hypothesis by predicting how changes in the independent variable will affect the dependent variable. Use the following format: **If** (an independent variable chosen from Question 3) increases/decreases, **then** (a dependent variable selected from Question 4) will increase/decrease/remain the same. If this is your first experiment, you should design an experiment with only one independent variable and one dependent variable. Later, you will learn how to design more complex experiments with multiple independent and dependent variables.

Notice how two different experiments can be designed using different responses to Questions 3 and 4.

Experiment 1	
Independent variable	Amount of fertilizer (increments of 5 g)
Dependent variable	Height of plants
Constants	Except for amount of fertilizer, all the potential variables listed as responses to Question 3 become the constants for this experiment.
Experiment 2	
Independent variable	Amount of water (increments of 50 ml)
Dependent variable	Number of leaves
Constants	Except for amount of water, all the potential variables listed as responses to Question 3 become the constants for this experiment.

In Experiment 1, the levels of the independent variable might be 0 g, 5 g, 10 g, and 15 g of fertilizer. Repeated trials would be 30 identical plants for each amount of fertilizer. The 30 plants that receive 0 g of fertilizer serve as a standard of comparison, or **control group**, for the experiment. The growth of the plants without fertilizer helps you understand how different amounts of fertilizer affects the height of other similar plants. For Experiment 2, what would you use as the control? How many repeated trials would you use? Why?

APPLYING THE FOUR QUESTION STRATEGY

You should now be ready to practice designing experiments on your own. Using the general topic of **motors**, work with a small group of classmates to brainstorm responses to the following four questions.

Q1: What materials are readily available for conducting experiments on (motors) ?
Q2: How do (motors) act?
Q3: How can I change the set of (motor) materials to affect the action?
Q4: How can I measure or describe the response of (motors) to the change?

After you have brainstormed your responses to these four questions, compare your responses to the following. How are yours similar? Different? For information about safety precautions when using electricity, see Section C in ***Experimenting Safely*** in front matter.

Q1: What materials are readily available for conducting experiments on (motors) ?
Hobby motors
Batteries
String
Weights

Q2: How do (motors) act?
Motors lift weights
Motors turn

Q3: How can I change the set of (motors) materials to affect the action?
Batteries
Brand
Voltage
Age
and so on

Hobby motors
Brand
Size
Shape
and so on

String
Length
Diameter
Type
and so on

Q4: How can I measure or describe the response of (motors) to the change?
Speed of lift
Amount of weight lifted
Number of times lifted

PROPS FOR BRAINSTORMING IDEAS FOR EXPERIMENTS

Ideas for experiments are all around you. Consider the following:

- simple and available materials listed in activities in textbooks and lab manuals;
- your hobbies, part-time jobs, or chores;
- science demonstration and "tricks" found in science activity books located in the children's science section of libraries and bookstores;
- "what if" questions, brief news summaries, and articles that suggest interesting follow up investigations.

Scientists do experiments to learn more about topics familiar to them. You can do the same. Start with lists, questions, and brief articles related to familiar objects, events, organisms, or the science subject you are studying. A list of materials, containing such items as various motor oils, a graduated cylinder, a balance, and squeaky wheels, might suggest some interesting experiments if you are studying the physical sciences. If you are a biology student, your thinking might be sparked by a list that includes beetles, grain, insecticides, and boxes. Questions involving familiar objects or pets, such as "What shapes do pets notice most?" might prompt a topic for investigation. Hobbies are a particularly good source of ideas for experiments. Use the *Four Question Strategy* to brainstorm ways you could vary the materials associated with one of your hobbies. Turn your hobby into a scientific investigation!

Books of science demonstrations, activities, and "tricks" can be great sources of ideas for experiments. Because they provide a list of necessary materials and a description of what action will occur, these books already answer Questions 1 and 2 of the *Four Question Strategy*. Select something from one of these books and answer Questions 3 and 4 to generate original experiments of your own. Consider the following science demonstration.

> **Going Over the Edge**
>
> **Materials:** water, pennies, and a glass
> **Procedure:**
>
> 1. Place the glass on a level table or other surface.
> 2. Fill the glass to the rim with water.
> 3. Hold the penny so that it is perpendicular to the surface of the water.
> 4. Slowly, lower the penny into the water and release it.
> 5. Repeat step 4 until the water spills over the glass.
>
> How many pennies did you add until the surface tension of the water was not able to keep the water from spilling over the rim?

Q1: What materials are readily available for conducting experiments on _(surface tension)_?
Glass
Pennies
Water

Q2: How does _(surface tension)_ act?
Forms a mound of water above the rim

Q3: How can I change the set of _(surface tension)_ materials to affect the action?

Glass	***Pennies***	***Water***
Size	Size	Size
Shape	*Age*	*Type*
Composition	*Cleanliness*	*Additives*
Height	*Other coins*	*Other liquids*
and so on	*and so on*	*and so on*

Q4: How can I measure or describe the response of _(surface tension)_ to the change?
Height of mound
Number of pennies to cause a spill

Stuck for an idea for an experiment? Try one of those science activity books. Other ideas can be found with the claims made in advertisements for all kinds of products. Designing a test for such claims can make a good experiment. Does that special plate really defrost foods faster? What *does* affect defrosting time?

Have you thought about cooking as a source of ideas for experiments? Does overbeating the batter for brownies really make a difference? Baking soda and baking powder produce carbon dioxide bubbles that make cakes rise. What if you increased or decreased the amount listed in the recipe? In baking bread there are specific directions about time, temperature, kneading, and yeast. Now, just suppose that you changed . . . Experiments are everywhere!

PRACTICING YOUR SKILLS

You know that the shortest distance between two points is a straight line. You also realize that brainstorming is not the fastest route to generating a topic to investigate. It is difficult to resist the temptation to list the first independent and dependent variables that come to mind, a few constants, a control, some repeated trials, and to yell out, "I'm done!" If you give in to this temptation, there will be two major losses. First, if you consider only a few variables, you may overlook a large number of factors that should be held constant. If you do not hold these constant, you will not have a fair test of your independent variable. The second loss is worse. Too rapid a choice of an experiment prevents you from being aware of numerous other variables to investigate that are potentially far more interesting than one produced by a quick thought. You will miss the intellectual excitement that results from personally discovering all sorts of possibilities.

EVALUATING YOUR RESPONSES TO THE FOUR QUESTIONS

How will you know if you have effectively used the *Four Question Strategy*? Checklists, such as Table 3.1, *Checklist for Evaluating Responses to Four Questions*, can help you evaluate your responses.

Q1: What materials are readily available for conducting experiments on ________________ ________________?
Did you produce an excellent, good, or poor list? That is a judgement call, but an excellent response should list all the materials needed to investigate your topic.

Q2: How do ________________ act?
To determine if your responses to Question 2 are correct, read the appropriate sections of textbooks, encyclopedias, or other reference materials and revise your response as necessary.

Q3: How can I change the set of ____________ materials to affect the action?
An excellent list would list numerous ways to vary each material. If you only listed a few ways to change a material, try to think of

TABLE 3.1 Checklist for Evaluating Responses to Four Questions

Criteria	Self	Peer/Family	Teacher
Q. 1: Readily available materials			
Excellent list Good list Poor list			
Q. 2: Action of materials			
Excellent answer (correct) Good answer (partially correct) Poor answer (incorrect)			
Q. 3: Ways to vary materials			
Excellent list Good list Poor list			
Q. 4: Ways to measure actions			
Excellent list Good list Poor list			
Creativity of topic			
Creativity of brainstorming			

other ways; ask your classmates or parents for their ideas. If you forgot one of the materials, add it now and think of ways to vary that material.

Q4: How can I measure or describe the response of ______________ to the change?
Look back at the actions you listed for Question 2. An excellent response to Question 4 would be several different ways to measure each action you previously listed.

Helping you develop the skills needed for producing a well-designed experiment is only one of the benefits of using the *Four Question Strategy*. There is also great satisfaction in achieving the scientific value of ***longing to know and understand.*** Finding the answer to "I wonder what would happen if I . . ." makes the effort worthwhile.

Related Web Sites

http://sln.fi.edu/tfi/activity/act-summ.html
http://www.ars.usda.gov/is/kids/fair/ideas.htm
http://members.aol.com/ScienzFairs/ideas.htm
http://www.isd77.K12.mn.us/resources/cf/SciProjIntro.html
http://ibms50.scri.fsu.edu/~dennisl/CMS.html
http://www.ed.gov/pubs/parents/Science
http://www.mcrel.org/resources/links/index.asp
http://www.awesomelibrary.org/science.html
http://nyelabs.kcts.org/flash_go.html
http://www.scri.fsu.edu/~dennisl/CMS/special/sf-hints.html (basic hints)
http://134.121.112.29/sciforum/guiding.html (Questions as prompts)
http://www.isd77.k12.mn.us/resources/cf/SciProjInter.html (general discussion of experimenting)
http://www.eduzone.com/Tips/science/SHOWTIP2.HTM
http://www.stemnet.nf.ca/~jbarron/scifair.html
http://www.sci.mus.mn.us/sln/tf/nav/thinkingfountain.html
http://www.exploratorium.edu/learning_studio/index.html
http://kidscience.miningco.com
http://www.waterw.com/~science/sample.html
http://weber.u.washington.edu/~chudler/experi.html (Human Biology)
http://www.flash.net/~spartech/ReekoScience/ReekoIndex.htm
http://ericir.syr.edu/Projects/Newton
http://www.eecs.umich.edu/mathscience/funexperiments/agesubject/age.html
http://www.eskimo.com/~billb/amasci.html
http://youth.net/nsrc/sci/sci.001.html
http://www.ksw.org.uk/
physics/1_curric/curric.html

Practice

1. Use the *Four Question Strategy* to brainstorm ideas for experiments on the following topics as designated by your teacher. Save your answers to this question for use in Chapter 4, Question 2.

   A. Meal worms
   B. Insect repellent
   C. Molds
   D. Disinfectants
   E. Magnets
   F. Furniture polish
   G. Glass cleaners
   H. Bathroom cleaners
   I. Sodas
   J. Insulation
   K. Heart rate
   L. Dishwashing liquid
   M. White socks/household bleach
   N. Ice cubes
   O. Paints
   P. Kitty litter
   Q. Carpet fibers
   R. Bread dough
   S. Car wax
   T. Denture cleansers

2. Using one of the following suggested lists of materials, brainstorm ideas for an experiment.

List A	List B	List C	List D
glasses	freezer	two varieties fruit flies	peat moss
water	ice cube trays	heat source	wood and metal blocks
identical coins	salt	light source	newspaper
soaps	juices	containers	soils
hot plate	food coloring	fruit and vegetables	sand
salts	rubbing alcohol	insecticides	water
	construction paper		dilute acid
	containers		fruit
			large flat containers

3. Use one of the following hypothetical newspaper stories as a basis for brainstorming an experiment.

   Pier Point City: Fishing guide, Captain Joe Finn, reported that the fishing season is in full swing. When asked about the best way to fish, he replied, "Time is important; the hours just after sunrise and just before sunset are best. By all means, use red colored artificial lures and six inch plastic worms."

   Fiberville: Ms. A. Boss, plant manager of The Fibers Research Division, reports that braiding the cotton threads produces a stronger twine than twisting them or using them as three straight strands. She also reported that soaking the threads in starch or glue solutions before making the twine has no effect on the strength of the twine.

   Duddville: Dr. I. M. Smart issued a report today that is highly critical of U.S. education. He claims that today's students are less knowledgeable than their parents. He also claims that less than 30% of high school seniors can name four or more cabinet members. He said that most students cannot identify more than 30% of the countries on a world map. He further stated that the high school seniors are less able than their parents to successfully solve problems involving percentages or fractions.

CHAPTER 4

Describing Experimental Procedures

Objectives

- Write clear and precise experimental procedures.
- Use the *Four Question Strategy* to assist you in writing clear and precise procedures.
- Use a checklist to evaluate procedures and identify needed improvements.

National Standards Connections

- Communicate scientific procedures and explanations (NSES).

Does the following conversation sound familiar between a science teacher and a student named George?

"George, in your procedure for your experiment there is a step missing between placing the acid in the beaker and beginning to add the base from the burette." "Yeah," says he upon reflection, "I put the indicator phen . . . phenol . . . phenolth . . ."

"Phenolthalein," the teacher says.

"Yeah, that stuff. I put it in with the acid," says George.

"Well, you didn't say so here!"

"No, but that's the point of the experiment. Everybody knows that. I added it, didn't I? If I hadn't, I'd still be adding base. I know, you know, everybody knows you must add an indicator at that point, so I just skipped writing the obvious."

FROM BRAINSTORMED IDEAS TO EXPERIMENTAL DESIGN TO PROCEDURE

You probably know someone like George. Writing a detailed procedure for a scientific investigation is a major challenge, but it must be done.

When details, however obvious, are left out of a procedure, readers will substitute their own way of doing the omitted step. Variations that appear to be extremely minor can have **drastic effects on the results of an experiment** and that is the problem. Some things are so obvious to writers that they fail to write them down. The same things are not obvious to everyone else. What is needed is a strategy that will allow you to state a procedure so clearly that everyone using it will do the same thing. For practice, begin with a familiar task, such as popping popcorn. Use the *Four Question Strategy* to brainstorm ideas about popping corn the "old fashioned way," without a microwave.

BRAINSTORMING IDEAS

Q1: What materials are readily available for conducting experiments on (**popcorn**)?
popcorn
oil
popper

Q2: How does (**popcorn**) act?
It pops.

Q3: How can you change the set of (**popcorn**) materials to affect the action?

Popcorn
Brand
Amount
Age

Oil
Brand
Amount
Kind

Popper
Brand
Heating time
Cooking time

Q4: How can you measure or describe the response of (**popcorn**) to the change?
Count the number of popped kernels
Measure the mass of the popped and unpopped kernels
Describe the appearance of the popcorn, e.g., color, shape of popped kernels

CLARIFYING EXPERIMENTAL COMPONENTS

Your responses to the four questions identify the independent variable, the dependent variable, and the constants. Use these components as clues to identify the other parts needed to complete the design of an experiment.

Independent Variable

Begin by identifying an independent variable. An independent variable is the variable that the experimenter changes on purpose. The responses to Question 3 are a list of potential ways to change the set of popcorn materials. Which one should be selected? It's up to you, the experimenter, but you should choose only one. In a simple experiment there is only one independent variable. Suppose you selected the amount of oil as your independent variable. That choice means that you will vary the amounts of oil placed with each batch of popcorn.

Levels of the Independent Variable

Next, the experimenter must decide on the levels of the independent variable. In this experiment, the levels of the independent variable are the amounts of oil tested. Four levels, such as 0, 10, 20, and 30 ml of oil were used, but you could also use just two levels, such as 0 and 30 ml of oil. Decisions about the number of levels and the value of each level must be made. With too few levels of the independent variable there will not be enough data points to clearly establish a pattern in the data. Too many levels give more data than the experimenter needs. Before making these decisions, cookbooks and library resources should be used to learn as much as possible about the variables, oil and popping corn.

Control

If oil is added to each batch of popcorn in an experiment and all the batches pop reasonably well, how would the experimenter know what difference the oil made? Maybe the corn popped well for other reasons. Maybe it might have popped even better with no oil. To determine what effect the oil had, a comparison group or control is needed. In this experiment, the control is the batch of popcorn that received no oil (0 ml).

Dependent Variable

In what ways might varying the amount of oil affect the popcorn? In Question 4, several possibilities were generated. Any one of them could be the dependent variable. The experimenter must select one, such as the number of kernels popped. That option becomes the dependent variable for the experiment. The experimenter must also decide how to clearly define a popped kernel. Will the experimenter count kernels that are partially popped or only those that are fully popped? How will each experimenter know the difference?

Hypothesis

Now it is time to write a hypothesis. A hypothesis is a prediction of the effect that changes in the independent variable will have on the dependent variable. One possible hypothesis would be: *If I increase the amount of oil put in the popper, then the number of popped kernels will increase.*

Constants

If only one independent variable is needed for an experiment, why was it important to brainstorm as many responses to Question 3 as possible? Each response to Question 3 is a potential independent variable. If the experimenter chooses the amount of oil as the independent variable, then all other potential independent variables listed under popcorn, oil, and popper must be made constants. To change a potential independent variable into a constant, assign it a specific value—amount, brand, time, and so on.

Popcorn	**Value**
Brand	"Pop-Rite Corn"
Amount	100 kernels
Age	1 year old
Storage	Air-tight container
Oil	
Brand	"Pazol"
Kind	Corn
Substitute	None
Popper	
Brand	"Cor-Pop"
Heating Time	2 min.
Cooking Time	4 min.

When the experimenter assigns a value for each constant, it is important to assign the value carefully. If the value of some of the constants are too high or too low the constants can interfere with the effects of the independent variable on the dependent variable. Suppose, for methods of storage, the experimenter assigned the value "in an open jar" rather than "in an air-tight container." The improper storage may have changed the popcorn so that it will not cook properly under any conditions. If so, it would not be possible to determine how various amounts of oil affect the popcorn because of the overwhelming effect of improper storage. To make good choices for constants, you will have to learn as much as possible about the topic of your experiment. Consult textbooks, library resources, and people who live in your community.

Repeated Trials

As an experimenter, you will also need to decide how many repeated trials are needed for each level of the independent variable. How many times will popcorn be popped with each different amount of oil—3, 6, 10, or more times? Besides the very practical considerations of time and expense, there are two other considerations that help to define the number of repeated trials that should be conducted in an experiment. These considerations are the amount of variation in the set of organisms or objects being used and the consequences of coming to a wrong conclusion. There is apt to be more variation in popcorn kernels than there is in the measurement of oil. Therefore, an experiment in which the kind of popcorn is the independent variable would need more trials than an experiment involving different amounts of oil.

The consequences of doing an experiment to determine which brand of popcorn to buy for a movie theater chain is greater than doing it to determine which kind to buy for your personal use. Thus, an experimenter would use a larger number of repeated trials in testing for the theater chain. The more trials used for each level of the independent variable, the more confident the experimenter can be in the results.

EXPERIMENTAL DESIGN

A verbal description of an experiment can be very long. Use an **experimental design diagram** as a concise way to describe an entire experiment. Look at the experimental design diagram in Figure 4.1. Does it communicate all the basic parts of the experiment on oil and popcorn? If not, what improvements would you make?

Now that all the major components of this experiment are identified, it is time to write a formal procedure. Start by closing your eyes and visualizing the steps you would follow when popping corn. First, list the main steps on paper. How does your list compare with the one below?

Step 1 Obtain popper
Step 2 Measure popcorn
Step 3 Measure oil
Step 4 Put oil in popper
Step 5 Heat popper
Step 6 Add popcorn
Step 7 Cook

In some instances, such as in Steps 1 to 3, the sequence is not important. In other instances, such as in Steps 4 to 7, the sequence is critical. Modify the list to include the specific independent variable and constants identified from Question 3 and the specific response of the dependent variable that is to be measured (Question 4).

Step 1	Obtain popper	"Cor-Pop"
Step 2	Measure popcorn	"Pop-Rite Corn" 100 kernels
Step 3	Measure oil	0, 10, 20, 30 ml
Step 4	Put oil in popper	
Step 5	Heat popper	2 minutes
Step 6	Add popcorn	100 kernels
Step 7	Cook	4 minutes
Step 8	Measure	Number of popped kernels

When you compare the list in Question 3 with the visualized steps in popping corn, you may notice that important variables, constants, and materials were omitted. For example, an instrument for measuring the oil, a graduated cylin-

Title: The Effect of Various Amounts of Oil on the Number of Popped Corn Kernels
Hypothesis: If I increase the amount of oil placed on the popcorn, then the number of popped kernels will increase.

IV: Amount of oil (ml)			
0% (Control)	10%	20%	30%
3 Trials	3 Trials	3 Trials	3 Trials

DV: Number of popped kernels
C: *Popcorn* "Pop-Rite" corn, 100 kernels, fresh
Oil "Pazol", Corn
Popper "Cor-Pop" brand, heating time of 2 minutes, cooking time of 4 minutes.

Figure 4.1 Experimental Design Diagram.

der, should be included. Using the lists as a guide, write a step-by-step procedure. How does your list of steps compare with the steps below?

Step 1 Obtain one "Cor-Pop" popcorn popper.
Step 2 Count out 100 kernels of fresh "Pop-Rite" corn.
Step 3 Measure 10 ml of "Pazol" oil with a graduated cylinder.
Step 4 Put the oil in the popper.
Step 5 Heat the oil for 2 min.
Step 6 Add the popcorn.
Step 7 Cook the popcorn for 4 min.
Step 8 Count the number of popped kernels.
Step 9 Repeat Step 1–8 for three trials.
Step 10 Repeat Steps 1–9 for the other amounts of oil e.g., 0, 20, 30 ml.

Note that this list consists of short complete sentences. You may find it helpful to first write the procedure for one value of the independent variable, such as 10 ml of oil. Next, add a step that tells the reader to repeat all the previous steps "x" number of times for the appropriate number of repeated trials (Step 9). Finally, add another statement to repeat the steps for additional values of the independent variable (Step 10). Do not write Steps 1–9 for 0 ml, as it does not make sense to measure 0 ml of oil to put in a popper. You may have noted a missing step. As a safety precaution, the cooked popcorn (Step 7) should be poured on a surface to cool before counting the number of popped kernels (Step 8). For more information about safety precautions, see ***Experimenting Safely*** in front matter.

If you are entering your experiment in a competition, check the competition's rules about procedure writing. Some competitions allow a procedure to be written as lists of steps and materials. Other competitions require the procedure to be written in paragraph form. If the competition requires a paragraph, it is probably easier for you to list the materials and steps of a procedure first and then change them into a paragraph. For example, the procedure for the experiment on the effect of various amounts of oil on the number of popped kernels could be written as a paragraph.

> Ten ml of "Pazol" oil were placed in a "Cor-Pop" popcorn popper and heated for 2 minutes. One hundred kernels of "Pop-Rite" popcorn were added to the popper. The popcorn was cooked for 4 minutes. After pouring the corn on a surface and allowing it to cool, the number of fully popped kernels was counted. After allowing the popper to cool and be cleaned, the procedure was repeated for a total of 3 trials for each amount of oil, 0 ml, 10 ml, 20 ml, and 30 ml.

EVALUATING YOUR PROCEDURES

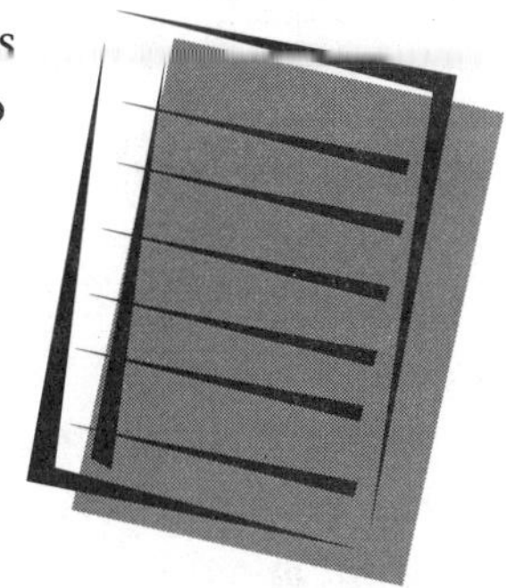

Writing procedures in this chapter began with two skills you already had—using the *Four Question Strategy* to brainstorm responses and drawing an experimental design diagram. To check your skills in these areas, continue to use the checklists in Chapters 2 and 3.

To check your new skill, writing a list of steps or a paragraph of procedures for an experiment, use the criteria in Table 4.1 *Checklist for Evaluating Procedures*. Remember to ask your friends, family, and teacher for comments. With their help, you can write a procedure so clear that everyone using it will do the same thing.

PRACTICING YOUR SKILLS

Now that you have learned about popping corn the "old fashioned way," substitute a microwave oven for the popper. Brainstorm a new set of responses using the *Four Question Strategy*. Decide on an experiment you might conduct, draw an experimental design diagram, and write a procedure.

For additional practice, refer to any of the brainstormed lists in Chapter 3. Select indepen-

TABLE 4.1 Checklist for Evaluating Procedures

Criteria	Self	Peer/Family	Teacher
All steps included			
All materials/equipment included			
Written for one level of independent variable			
Repetitions for repeated trials			
Repetitions for levels of independent variable			
Written in approved format—lists or paragraph			
Spelling/Grammar			
Sentence/Paragraph structure			

dent and dependent variables that interest you. Draw an experimental design diagram and write a procedure for the experiment.

When you have demonstrated skill in writing procedures for hypothetical experiments, you are ready to write procedures for your own experiment. When you first write your procedures, view them as a first draft. Form a work group with your classmates and analyze each others' procedures and revise as necessary. As a final strategy, you may test the completeness of your procedures by acting out exactly what you wrote.

Related Web Sites

http://www.isd77.K12.mn.us/resources/cf/SciProjIntro.html

Practice

1. Write a detailed and precise procedure that includes both the sequence of steps to be taken and the materials needed for each of the following activities that is assigned by your teacher.

   A. You want to find out how fast the temperature of 50 ml of luke warm water would rise if it were heated by a candle.
   B. Radio station WPIG is giving away $1,000 to the first caller to get through after they hear The Pig Squeal.
   C. Your dog has fleas and you need to evaluate the effectiveness of several different brands of flea collars.
   D. You need to make a twin bed with fresh linen.
   E. You need to wax a brand new Super Hawk auto.

2. Using your answers to the *Four Question Strategy* (see practice problems from Chapter 3), draw an experimental design diagram for *one* experiment you could conduct and write a procedure for the experiment you developed.

CHAPTER 5

Constructing Tables and Graphs

Objectives

- Construct an appropriate data table for organizing data.
- Construct a graph from a brief description of an investigation and a set of data.
- Draw a line-of-best-fit for experimental data.
- Describe the relationship between variables depicted on a graph.
- Identify data for which a bar graph or a line graph is most appropriate.
- Use a checklist to evaluate tables of data and graphs and identify needed improvements.

National Standards Connections

- Use appropriate tools and techniques to gather, analyze, and interpret data (NSES).
- Use technology and mathematics to improve investigations and communications (NSES).
- Construct and draw inferences from charts, tables, and graphs that summarizes data from real-world situations (NCTM).

"Jack, where are the data you have been collecting all period?"

"Here."

And you are shown 53 unlabeled numbers in groups of three or four, on both sides of the page.

"How do you know which numbers are for our independent variable and which are for our dependent variable?"

"Well," says Jack, "The first trial is up here in the upper left, then the next trial is to the right, or is it this set below the first?" "Oops," says he as he drops the paper and slowly picks it up. "Now let's see, is this the front of the sheet? Oh, never mind, I'll sort it out later. Don't worry!"

Have you ever had a lab partner like Jack? Organization and communication skills are definitely not Jack's strengths. Unlike Jack, scientists recognize that organizing data from experiments into tables and graphs helps them better understand their results. Organizing data into data tables and graphs is a lot like setting the table for dinner. Both require a plan for organizing—on which side of the plate does the fork go? Is the independent variable always on the left side of a data table and on the X axis of a graph? Guidelines for constructing tables of data, bar graphs, and line graphs are described in this chapter.

CONSTRUCTING DATA TABLES

Have you ever wondered if one brand of *paper towel* is really a 'quicker picker upper' than another? Do you think commercials comparing brands are fair tests of which brand is best? What about constants and repeated trials? Could *you* conduct a fair test of different brands? Design an investigation to determine the effect of submersion time of paper towels on liquid absorption or use the procedure outlined in Investigation 5.1, *Time and Absorption.* Before you begin collecting your data, how would you record it? Use the following questions to evaluate your method of recording data.

- Does your system communicate the relationship between the independent and dependent variables?
- Does your system communicate the order in which the independent variable was changed?
- Does your system's title communicate the purpose of the experiment?

Although there are no absolute rules for constructing data tables, there are generally accepted guidelines (see Table 5.1 *The Effect of Submersion Time on the Height a Liquid Rose in a Paper Towel*). For example, the independent variable is almost always recorded in the left column and the dependent variable in the right. When repeated trials are conducted, they are recorded in subdivisions of the dependent variable column. If derived quantities (such as the average height risen per second of submersion) are calculated, they are recorded in an additional column to the right. When recording data in a table, the values of the independent variable are ordered. The data may be arranged from smallest to largest or from largest to smallest. Although no rule exists, most data are ordered from smallest to largest. The title of the data table should clearly communicate the purpose of the experiment through specific references to the variables under investigation. For example, see Table 5.1 *The Effect of Submersion Time on the Height a Liquid Rose in a Paper Towel.* Table 5.1 also illustrates how these guidelines can be used to construct a table of data.

TABLE 5.1 The Effect of Submersion Time on the Height a Liquid Rose in a Paper Towel

Column for independent variable	***Column for dependent variable***			***Column for derived quantity****
	Height liquid rose in towel (mm) **Trials**			
Time paper towel submerged (sec)	**1**	**2**	**3**	**Average height (mm)**
10	11	10	11	11
15	14	14	13	14
20	14	14	14	14
25	15	15	16	15
30	16	16	16	16
35	17	17	18	17
40	19	20	19	19

*In formal data tables, the information in *ITALIC* is not included.

TURN TO Investigation 5.1, page 40

EVALUATING YOUR SKILLS WITH DATA TABLES

After you complete your table of data, rate it using the criteria listed in Table 5.2 *Checklist for Evaluating Data Tables.*

CONSTRUCTING LINE GRAPHS

Graphs communicate in pictorial form the data collected in an experiment. Usually, a well-constructed graph communicates experimental findings better than a data table. However, graphs are more difficult to construct and involve several subskills, including knowledge of the major parts of a graph, relating data pairs from a data table to data pairs on a graph, constructing an appropriate scale for each axis, plotting the data on a graph, and summarizing the trends through a line-of-best-fit and descriptive sentences. By using a series of steps, you can easily construct the graph you need to display your data. Begin by drawing and labeling the axes.

Draw and Label Axes

Graphs are pictorial representations in two dimensions—horizontal and vertical. Scientists place the independent variable, for example, time of submersion, on the X or horizontal axis and the dependent variable, for example, height the liquid rose, on the Y or vertical axis. The unit of measurement is placed in parentheses next to or beneath the variable (see Figure 5.1).

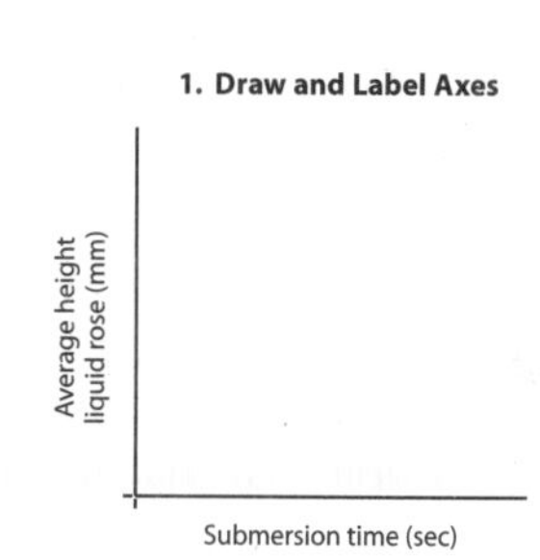

TABLE 5.2 Checklist for Evaluating Data Tables

Criteria	Self	Peer/Family	Teacher
Title			
Vertical column for independent variable			
Title/Unit of independent variable included			
Values of independent variable ordered			
Vertical column for dependent variable			
Title/Unit of dependent variable included			
DV column subdivided for repeated trials			
Values for DV correctly entered			
Vertical column for derived quantity			
Unit of derived quantity included			
Derived quantity correctly calculated			

INVESTIGATION 5.1 ▪ Time and Absorption

Materials

- Paper towel
- Food coloring
- Plastic cup or beaker
- Water
- Clock with second hand
- Scissors
- Metric ruler
- Pencil

Safety

- Handle sharp objects safely.
- Wash hands.
- Wear goggles.

Procedure

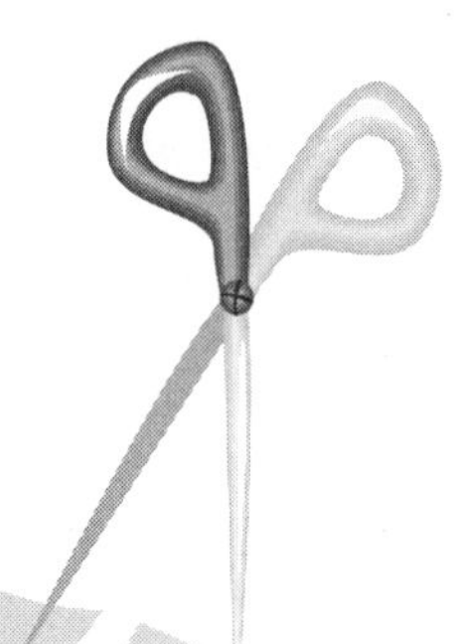

1. Cut a paper towel into strips, 2 cm x 22 cm.
2. Fill a container (cup or beaker) with water. Add several drops of food coloring.
3. Place the paper towel strip 1 cm into the colored water for the time interval designated by your teacher, for example, 10, 15, 20, 25, 30, 35, or 40 seconds.
4. At the end of each time interval, *quickly* mark the water levels with a pencil. Then, measure the height the liquid rose in mm and record the data.
5. Repeat Steps 1 through 4 two more times for a total of 3 trials.
6. Calculate the average height the liquid rose (mm).

Class Data Table

Construct a class data table using the following guidelines.

- Make a table containing vertical columns for the independent variable, dependent variable, and derived quantity (average height).
- Subdivide the column for the dependent variable to reflect the number of trials.
- Order the values of the independent variable—preferably from the smallest to the largest.
- Record the values of the dependent variable that correspond to each value of the independent variable.
- Calculate the derived quantities and enter the values into the table.

Column for independent variable	Column for dependent variable				Column for derived quantity
Time paper towel submerged (sec)	Height liquid rose in towel (mm) Trials				Average height (mm)
	1	2	3	etc.	

(continued on the following page)

INVESTIGATION 5.1 ■ Time and Absorption *(continued)*

Constructing a Line Graph

Follow the sequence described by your teacher to construct a line graph for the class data.

- Draw and label the X and Y axes of the graph.
- Write data pairs for the values of the independent and the dependent variable; use the derived quantity (average height) for the dependent variable.
- Determine an appropriate scale for the X axis and the Y axis.
- Plot the data pairs on the graph.
- Summarize the data trends with a line-of-best-fit and descriptive sentences.

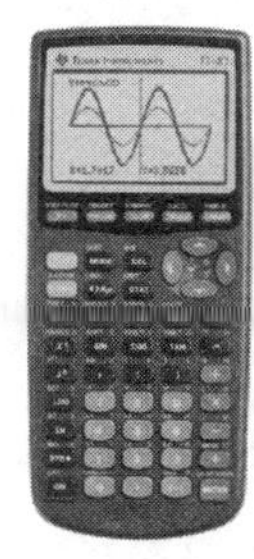

USING TECHNOLOGY

1. In the **STAT** mode of your calculator, enter the values for the time the paper towel was submerged in List 1 and values for the mean height the liquid rose in List 2. (See Appendix A, *Using Technology*, for additional help in using the graphing calculator.)
2. In setting up your graph, select scatter plot as your graph type and List 1 for your x values and List 2 for your y values. Graph the data.
3. Examine the trend of the data and calculate a line of best fit by performing the appropriate regression (equation) analysis. For example, if the general trend of the data appears to be straight, you may wish to calculate a linear regression. Copy the calculated values to an empty Y=, and graph the equation.
4. To predict height values for submersion times that were not measured, press **Trace** and then the arrow keys to move along the best fit line to see the predicted y value (height) for any desired x value (time submerged). Depending on the brand of your calculator you may have to adjust the 'Window' to predict or extrapolate beyond the minimum or maximum x values.

Write Data Pairs

2. Write Data Pairs

(10, 11)
(15, 14)
(20, 14)
(25, 15)
(30, 16)
(35, 17)
(40, 19)

Points on a graph are represented by a set of data. The value for the horizontal (X) axis is written first, followed by the corresponding value for the vertical (Y) axis. The two numbers are separated by a comma. Both values are placed in parentheses, for example, (10,11). In mathematics, data pairs are called *number pairs.* If the guideline for placement of the IV and DV in the data table is followed, the data pair will be in the same sequence (see Figure 5.1).

Example Scale for X axis: *submersion time (sec)*	
Largest value	40 sec
Smallest value	10 sec
Difference	30 sec
Difference divided by 5	30 sec ÷ 5 = 6 sec
Quotient rounded to counting number	6 sec rounded to 5 sec

Example Scale for Y axis: *average height liquid rose (mm)*	
Largest value	19 mm
Smallest value	11 mm
Difference	8 mm
Difference divided by 5	8 mm ÷ 5 = 1.6 mm
Quotient rounded to counting number	1.6 mm rounded to 2 mm

Determine Scales for Axes

The most challenging part of constructing graphs is determining the right scale for numbering the axes of a graph. An easy way to find a good scale to fit the data consists of a series of steps described below and summarized in Figure 5.1.

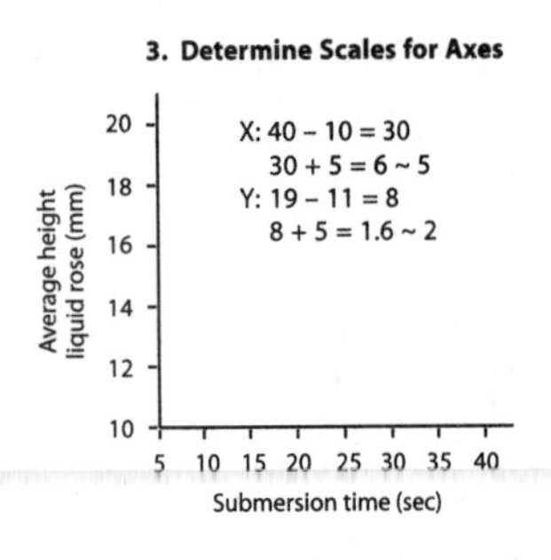

1. To determine an acceptable scale for the axis, find the range of the data to be graphed by finding the difference between the smallest and largest values for the variable. To determine the size of each interval, divide the range by 5. (Dividing by 5 results in 5–7 intervals, which is a reasonable number of intervals.) Too many intervals crowd a graph and too few make it difficult to plot points. After dividing the difference by 5, round the resulting quotient to the nearest convenient counting number. Any number that is easily counted in multiples works well, such as multiples of 2, 5, or 10.

2. Develop a scale for each axis using the rounded quotient as the interval. Begin with an interval that is less than the smallest value to be graphed and end with an interval that allows the largest value to be graphed. In Table 5.1, the shortest submersion time is 10 sec so the scale for the X axis would begin with 5 sec; the scale would end with the largest value, 40 sec. Because the smallest value of the dependent variable to be graphed is 11 mm, the scale for the Y axis would begin with 10 mm; the scale would end with 20 mm to allow the largest value, 19 mm, to be graphed. Some people think that all graphs begin with the intersection of the X and Y axes labeled as (0,0). Graphs that begin at (0,0) work fine when the data sets start at or near zero. But in many experiments, data sets begin with numbers far from zero, resulting in a large gap between 0 and the first piece of data to be graphed. If you are uneasy about not beginning at zero, ask your teacher about using the symbol ⫽ to indicate the part of the graph that is not shown.

Plot Data Pairs

Using a data pair such as (10, 11) from Table 5.1, plot this point by locating the 10 on the X axis

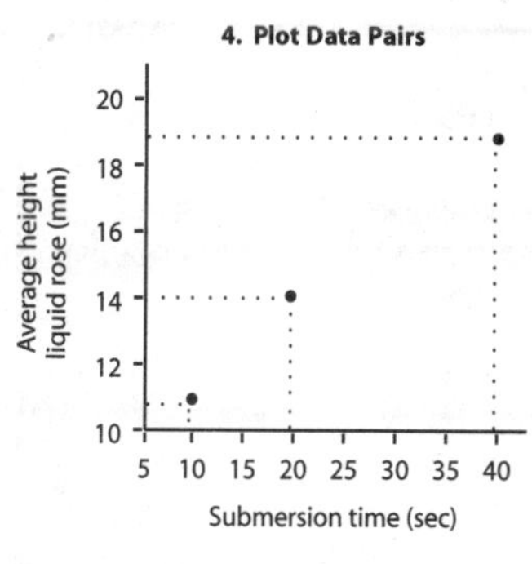

and the 11 on the Y axis. By sighting an imaginary line straight up from the 10 and another straight across from the 11, a point can be plotted where the two lines cross. As illustrated in Figure 5.1, use the same procedure to plot other data pairs, such as (20, 14) and (40, 19).

Summarize Trends

Because experimental data are subject to error, data points on a graph are not directly connected. Instead a line-of-best-fit is used to communicate the general data pattern. To construct a line-of-best-fit, draw a line about which an equal number of data points fall to either side.

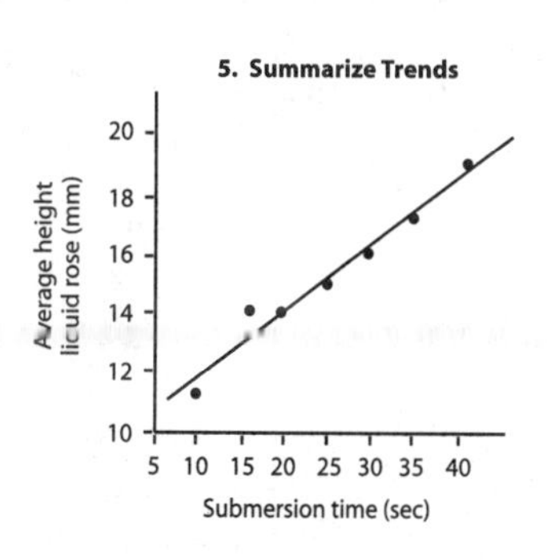

Some examples of Lines-of-Best-Fit follow.

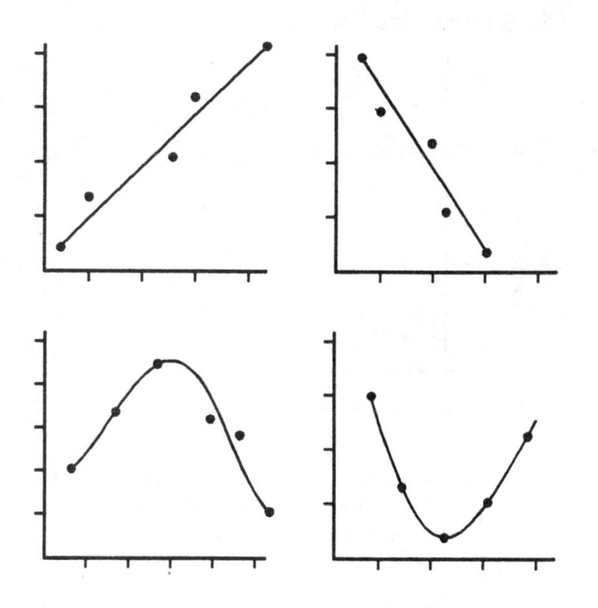

Describe the trend of the line-of-best-fit by writing a sentence that communicates what happens to the dependent variable as the independent variable is changed, e.g., *When the amount of time a paper towel was soaked increased, the height of the colored liquid also increased (see Figure 5.1).*

EVALUATING YOUR SKILLS WITH LINE GRAPHS

After you complete your line graph, rate it using the criteria listed in Table 5.3 *Checklist for Evaluating Line Graphs.*

BAR VERSUS LINE GRAPHS

Sometimes, you may not be sure whether to make a bar graph or a line graph of your data. The appropriate type of graph depends on the type of data collected. Observations and measurements of variables can be classified as either **discrete** or **continuous.** Discrete data are categorical or counted like days of the week, gender, kind of animal, brand of battery, number of children, or color. Bar graphs are appropriate for these types of variables. Other variables are continuous and associated with measurements involving a standard scale with equal intervals. Examples include the height of plants in centimeters, the amount of fertilizer in grams, and the length of time in seconds. When the data may be any value in a continuous range of measurements, a line graph is a better way to show the data. Line graphs allow you to infer the value of points on a graph that were not directly measured. There is an easy test for determining which type of graph is appropriate for a set of data. If the intervals between recorded data have meaning, a line graph is appropriate. When the intervals between the data do not have meaning, like product brands, a bar graph should be used to display the data.

CONSTRUCTING BAR GRAPHS

Because paper towel commercials try to convince consumers of the superior absorbing ability of their product, you may wonder which brand is superior. Design an investigation to determine the relative effectiveness of six different brands of

Experimental Data

Independent variable Submersion time (sec.)	*Dependent variable* Average height liquid rose (mm)
10	11
15	14
20	14
25	15
30	16
35	17
40	19

1. Draw and Label Axes

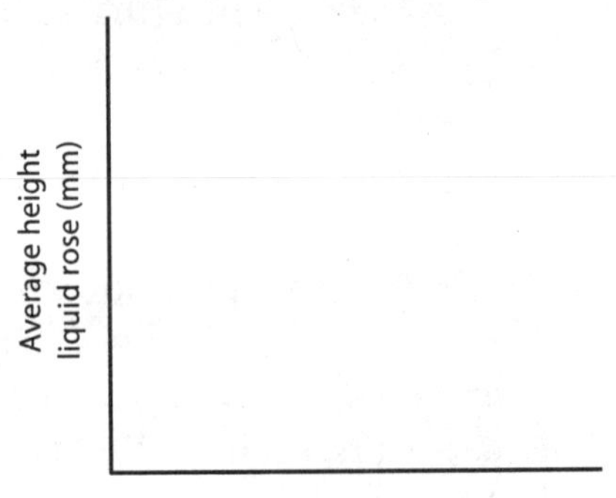

2. Write Data Pairs

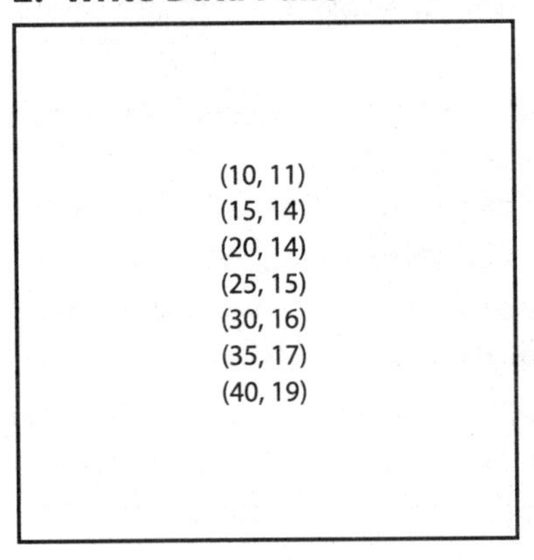

3. Determine Scales for Axes

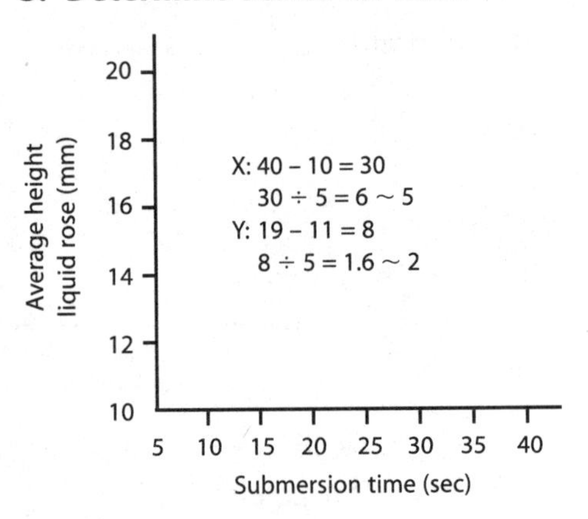

4. Plot Data Pairs

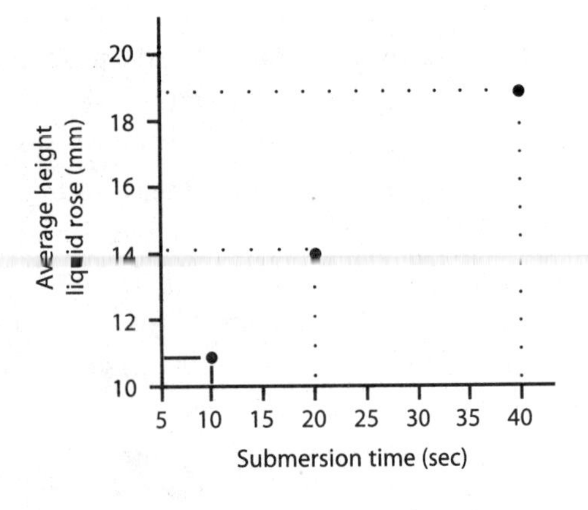

5. Summarize Trends

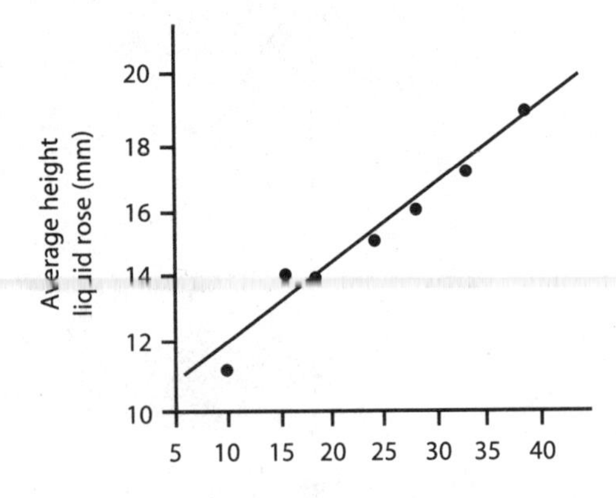

As the length of time the paper towel was submerged increased, the height the liquid rose also increased.

Figure 5.1 Constructing Line Graphs.

TABLE 5.3 Checklist for Evaluating Line Graphs

Criteria/Value (100 points)	Self	Peer/Family	Teacher
Title			
X axis correctly labeled including units			
Y axis correctly labeled including units			
X axis correctly subdivided into scale			
Y axis correctly subdivided into scale			
Data pairs correctly plotted			
Data trend summarized with line-of-best-fit			
Data trend summarized with sentences			

paper towels or use the procedure outlined in Investigation 5.2, *Brand and Absorption.* Before you begin the experiment, review the features of good data tables. Collect data quickly by working in groups to test different brands of paper towels.

After combining class data, you should distinguish between discrete data, like brands of paper towels, and continuous data, like the amount of water absorbed. A line graph of the data is not appropriate because intervals between the values of the independent variable (brand of paper towel) have no meaning. You should use a bar graph instead to display this data.

To construct a bar graph for the data from Investigation 5.2, *Brand and Absorption,* use a sequence of steps similar to those outlined for line graphs (see Figure 5.2).

- Draw and label the X axis (independent variable) and Y axis (dependent variable) of the graph.
- Write data pairs for the values of the independent and dependent variables.
- Subdivide the X axis to depict the discrete values of the independent variable; that is, the six paper towel brands. Evenly distribute the values along the axis leaving a space between each value.
- Determine an appropriate scale for the Y axis that depicts the continuous values of the dependent variable, water absorbed (ml); subdivide the Y axis.
- Draw a vertical bar from the value of the independent variable (X axis) to the corresponding value of the dependent variable (Y axis). Leave spaces between each bar.
- Summarize the graph with descriptive sentences.

TURN TO Investigation 5.2, page 48

Experimental Data

Independent variable Brand of paper towel	Dependent variable Water absorbed (ml)
A	34
B	17
C	24
D	36
E	27
F	25

1. Draw and Label Axes

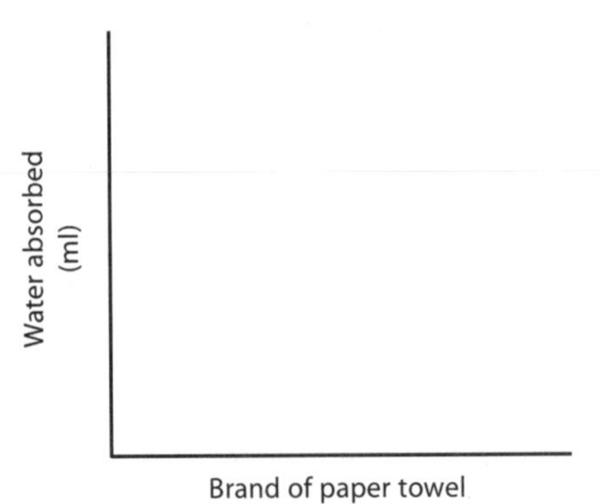

2. Write Data Pairs

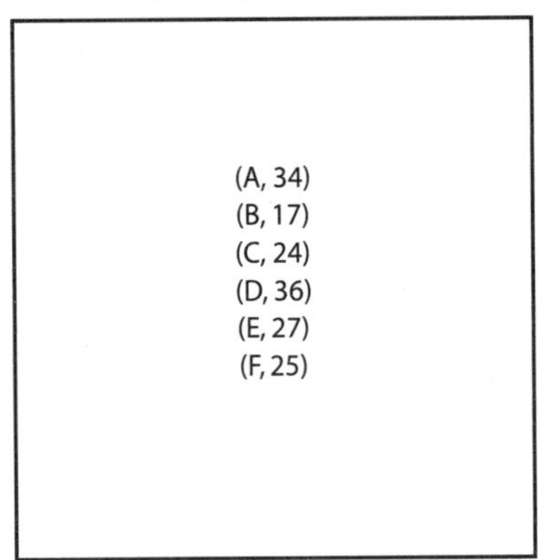

3. Determine Scales for Axes

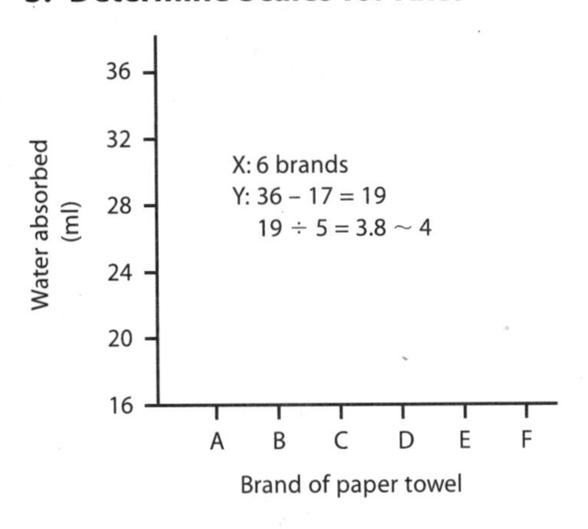

4. Plot Data Pairs

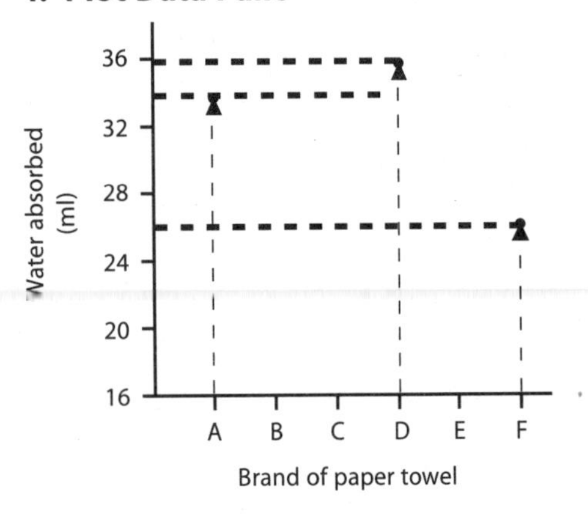

5. Summarize Trends

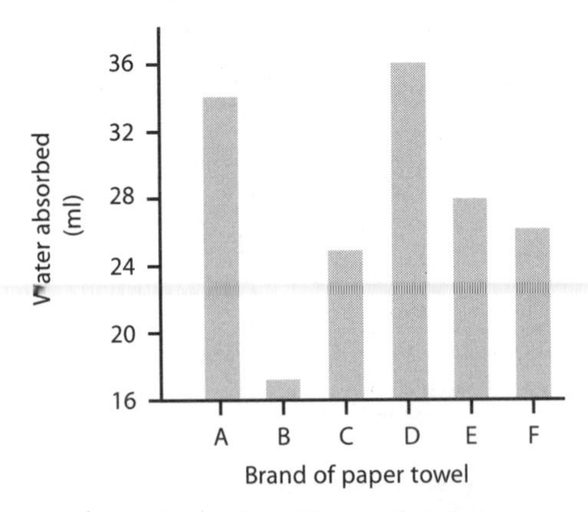

Brands A and D were the most effective water absorbers. The least effective absorber was Brand B. Brands C, E, and F absorbed intermediate amounts of water.

Figure 5.2 Constructing Bar Graphs.

EVALUATING YOUR SKILLS WITH BAR GRAPHS

After you complete your bar graph, rate it using the criteria listed in Table 5.4 *Checklist for Evaluating Bar Graphs.*

TABLE 5.4 Checklist for Evaluating Bar Graphs

Criteria/Value (100 points)	Self	Peer/Family	Teacher
Title			
X axis correctly labeled including units			
Y axis correctly labeled including units			
X axis correctly subdivided—discrete values			
Y axis correctly subdivided into scale			
Vertical bars for data pairs correctly drawn			
Data trend summarized with sentences			

REFERENCES

Rezba, R.J. et al. (1995). *Learning and assessing science process skills.* Dubuque, Iowa: Kendall/Hunt Publishing Company.

Gabel, D. (1993). *Introductory science skills.* Prospect Heights, Illinois: Waveland Press, Inc.

Related Web Sites

http://www.scri.fsu.edu/~dennisl/CMS/sf/sf_details.html

http://www.mste.uiuc.edu/stat/stat.html

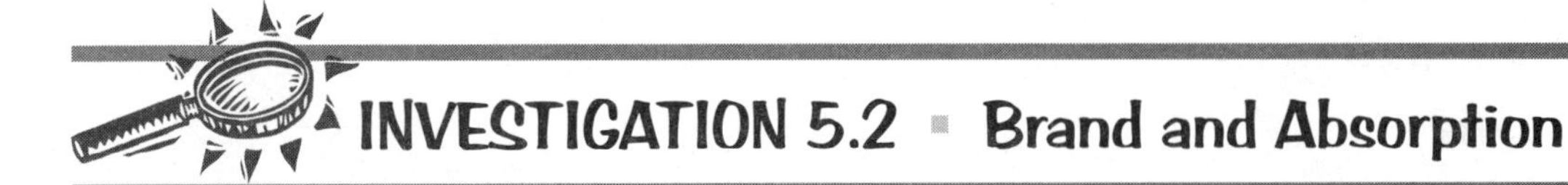

INVESTIGATION 5.2 ▪ Brand and Absorption

Materials

- Various brands of paper towels (A, B, C, D, E, F)
- Plastic container or 250 ml beaker
- Water
- Graduated cylinder (100 ml)
- Watch with second hand
- Pencil

Procedure

1. Measure 100 ml of water with the graduated cylinder. Add the water to the container (plastic container or 250 ml beaker).
2. Obtain one square of the paper towel brand designated by your teacher, such as, Brand A, B, C, D, E, or F.
3. Push the square of paper towel into the water for 30 sec. Use a pencil to push the towel under the surface.
4. Remove the paper towel. Hold the paper towel over the container until it stops dripping.
5. Use the graduated cylinder to measure the amount of water (ml) remaining in the container. Subtract the value from 100 ml to determine the amount of water (ml) absorbed by the towel.
6. Repeat Steps 1 to 5 for a total of 4 trials.
7. Calculate the average amount of liquid absorbed (ml).

Class Data Table

Construct a class data table using the following guidelines.

- Make a table containing vertical columns for the independent variable, dependent variable, and derived quantity (average water absorbed).
- Subdivide the column for the dependent variable to reflect the number of trials.
- Record the values of the independent variable (Brands A, B, C, D, E, and F).
- Record the values of the dependent variable that correspond to each value of the independent variable.
- Calculate the derived quantities and enter the values into the table.

Constructing a Bar Graph

Follow the sequence described by your teacher to construct a bar graph for the class data.

- Draw and label the X and Y axes of the graph.
- Write data pairs for the values of the independent and the dependent variables; use the derived quantities (average water absorbed) recorded in the class data table.

(continued on the following page)

INVESTIGATION 5.2 ▪ Brand and Absorption *(continued)*

- Subdivide the X axis to depict the discrete values of the independent variable, six paper towel brands. Evenly distribute the values along the axis, leaving a space between each value.
- Determine an appropriate scale for the Y axis that depicts the continuous values of the dependent variable, water absorbed (ml); subdivide the Y axis.
- Draw a vertical bar from the value of the independent variable on the X axis to the corresponding value of the dependent variable on the Y axis. Leave spaces between each bar.
- Summarize the data trends with descriptive sentences.

USING TECHNOLOGY

1. In the **STAT** mode of your calculator, enter consecutive numbers (e.g., 1, 2, 3 . . .) for the brands of paper towel in List 1 and the values for amount of water absorbed in List 2. (See Appendix A, *Using Technology*, for additional help in using the graphing calculator).
2. In setting up your graph, select a histogram (bar graph) as your graph type and List 1 for your x values and List 2 as the frequency. Graph the data. Depending on the brand of your calculator you may have the option to adjust the spacing and width of the bars.

Practice

1. For each experiment title listed, state whether the experiment should be graphed as a bar or a line graph.

   A. The Effect of Coloration on the Number of Kittens Sold at a Pet Store.
   B. The Effect of Concentration of Sugar Water on the Number of Visits of Hummingbirds to a Feeder.
   C. The Effectiveness of Different Brands of Paper Towels on the Absorption of Water.
   D. The Effect of the Horsepower of a Tractor on the Mass of a Sled It Can Pull.

2. Construct a data table and an appropriate graph for the following sets of data.

   A. Relationship Between Distance Below Surface and Number of Fossils Collected

Distance below surface (cm)	Number of fossils collected
80	0
140	2
200	8
260	15
320	32

   B. In 1988, the USA imported food. We imported in billions of dollars the following amounts: Shellfish, $2.7; Coffee, $2.5; Beef and Veal, $1.7; Pork, $0.9; Orange juice, $0.6; Cheese, $0.4; Grapes, $0.3; Tomatoes, $0.2.

   C. Bill and Sheri decided to study possible relationships involving abdominal muscle strength and endurance. Bill chose to see whether the number of situps that male athletes who are in top shape could do in two minutes is related to their ages. Sheri chose to see whether the weight of female athletes in top shape is related to the maximum number of situps they could do in two minutes. Construct data tables and graphs for Bill's and Sheri's data.

Bill's Data

Subject	Age (years)	Number of situps
Mark	14.5	95
Bob	17	100
Ron	15.5	97
Don	18	102
Lou	14	93
Armand	16.5	99
Norm	15	9
Bill	17.5	101
Doug	16	98

Sheri's Data

Subject	Weight (in pounds)	Number of situps
Gail	100	71
Margo	170	59
Joyce	180	57
Lori	110	69
Cathy	130	66
Dena	120	68
Tammy	150	62
Agnes	160	61
Linda	140	64

3. Describe the relationship between the variables graphed in Questions 2 A–C.

CHAPTER 6

Writing Simple Reports

Objectives

- Identify the elements of a simple report—title, introduction, experimental design diagram, procedure, results, and conclusion.
- Write a simple report of a scientific investigation.
- Use a checklist to evaluate a simple report and identify needed improvements.

National Standards Connections

- Develop descriptions, explanations, predictions, and models using evidence (NSES).
- Think critically and logically to make the relationship between evidence and explanations (NSES).
- Make inferences and convincing arguments that are based on data analysis (NCTM).

You can learn to write the critical elements of a formal scientific paper by first learning to write simple reports consisting of six components:

1. Title
2. Introduction
3. Experimental design diagram
4. Procedure
5. Results
6. Conclusion

In previous chapters you learned to write a title for an experiment, to draw an experimental design diagram, to write a clear procedure, and to show results through tables and graphs. You can write a simple report by combining these skills with two new skills—writing an introduction and a conclusion.

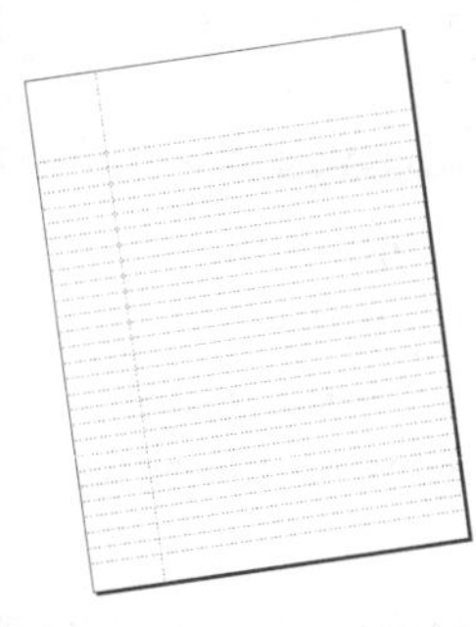

A good way to learn to write a simple report is to first collect some data from an experiment. Advertisements for household products suggest numerous questions for investigation. Think about advertisements for effervescent tablets that reduce stomach acid. Remember how they behave in water—plop, plop, fizz, fizz—do you think the temperature of the water would make a difference? Design your own investigation to test the hypothesis that if the tem-

INVESTIGATION 6.1 ▪ Super Fizzers

Directions

Read the list of materials and the procedure for conducting the investigation. Before you begin the experiment: (1) write a title, (2) draw an experimental design diagram, and (3) construct a data table. After completing the experiment, construct an appropriate graph. Refer to Tables 6.1 *How to Write a Simple Report* and 6.2 *Checklist for Evaluating a Simple Report* for examples.

Materials

- Plastic cups
- Water at three different temperatures (ice cold, room temperature, hot)
- Brand X effervescent tablets
- Watch with second hand
- Graduated cylinder
- Goggles

Safety

- Wear goggles.
- Handle hot objects carefully.

Procedure

1. From the central supply area, obtain 75 ml of ice water.
2. Add one effervescent tablet.
3. Record the time (sec) for the tablet to completely dissolve. Discard the solution as directed by your teacher.
4. From central supply, obtain 75 ml of room temperature water. Repeat Steps 2–3.
5. From central supply, obtain 75 ml of hot water. Repeat Steps 2–3.
6. To create repeated trials, record your group's data on the class data table.
7. Compute the average time for dissolving at each temperature using the values from the class data table.
8. Construct an appropriate graph of the data.

Writing a Simple Report

Follow the sequence described by your teacher to write a simple report for this investigation.

1. Title
2. Introduction
3. Experimental Design Diagram
4. Procedure
5. Results
6. Conclusion

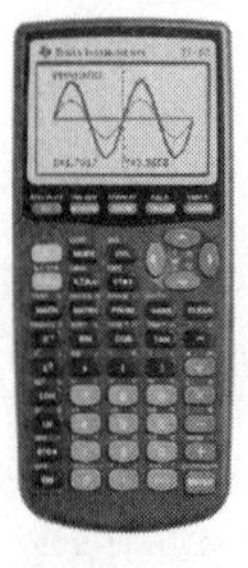

USING TECHNOLOGY

1. In the **STAT** mode of your calculator, enter consecutive numbers (e.g., 1, 2, 3) for the temperatures in List 1 and the values for mean dissolving time in List 2. (See Appendix A, *Using Technology*, for additional help in using the graphing calculator).
2. In setting up your graph, select a histogram (bar graph) as your graph type, List 1 for your x values (temperatures) and List 2 (dissolving time) as the frequency. Graph the data. Depending on the brand of your calculator you may have the option to adjust the spacing and width of the bars.

perature of the water is increased, **then** effervescent tablets will dissolve faster, or use the procedure outlined in Investigation 6.1, *Super Fizzers.* To save time and materials, you may wish to work in a group.

Use ice water, room temperature water, and hot water ~ 45°C as the cold, room, and hot water temperatures needed in the investigation. Before starting the investigation, you should write a title and a hypothesis, draw an experimental design diagram, and make a data table. After you have collected your data, construct an appropriate graph. Now only two parts of a report remain to be done—the introduction and the conclusion. You already know how to write four of the six components of a simple report. In the next section you will learn to write these two new parts and to combine all six components into a simple report.

TITLE

A good title relates the independent and dependent variables that were investigated. Initially, you may wish to write a title using a structured format: The Effect of the *Independent Variable* on the *Dependent Variable.*

Example:
The Effect of Water Temperature on the Dissolving Time of Effervescent Tablets

INTRODUCTION

In this part of a report, you describe your research problem by stating the rationale, purpose, and hypothesis for the study. An easy way to write an introduction is to write responses to each of the following questions.

Q1. Why did you conduct the experiment? *(Rationale)*
Some over-the-counter drugs are effervescent tablets that must be dissolved in water before drinking it. Directions, however, do not include any suggested water temperatures.

Q2. What did you hope to learn? *(Purpose)*
The purpose of this experiment was to determine the effect of water temperature on the time required for effervescent tablets to dissolve.

Q3. What did you think would happen? *(Hypothesis)*
The researcher hypothesized that if the temperature of the water is increased, then effervescent tablets will dissolve faster.

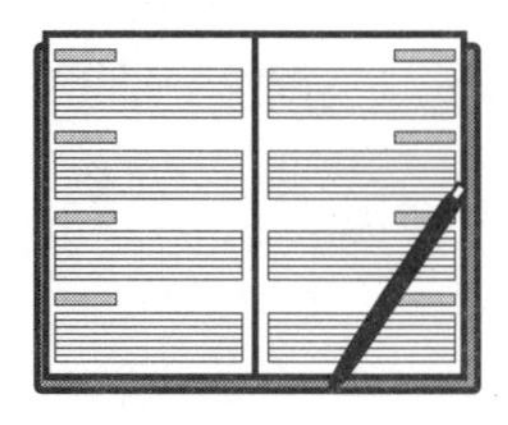

In the hypothesis, you communicate the expected effect of the independent variable on the dependent variable. A simple hypothesis may be written as an **if . . . then** statement, such as, "**If** soils contain a high percentage of clay (independent variable), **then** percolating rainwater will be more acidic (dependent variable)." An explanation of how you became interested in the study may also be included in the introduction section.

Notice that the responses to Questions 2 and 3 did not begin with "I wanted to know . . .," "I thought that if the temperature . . .," or "I hypothesized that . . ." Scientific writing is more formal and is written in third person (the investigator, the researchers) rather than in first person (I or we). Rather than say, "I hypothesized that . . .", you need to refer to yourself as the 'investigator' or the 'researcher' and write, "The investigator hypothesized that . . ."

EXPERIMENTAL DESIGN DIAGRAM

As you learned in Chapter 2, your experimental design can be described using a simple diagram. Remember to first draw a rectangle. The independent variable (IV) is written across the top of a rectangle that is divided into labeled columns to represent the different values of the independent variable. The number of trials is indicated in the columns. The dependent variables (DV) and constants (C) are written beneath the rectangle.

TABLE 6.1 How to Write a Simple Report

Part	Purpose
1. Title	Write a sentence that relates the independent and dependent variables that were investigated.
2. Introduction	Describe the rationale, purpose, and hypothesis for the investigation. Use three questions to guide your writing of the introduction. ■ Why did you conduct the experiment? (Rationale) ■ What did you hope to learn? (Purpose) ■ What did you think would happen? (Hypothesis)
3. Experimental Design Diagram	Format the experimental process. ■ Begin the diagram by drawing a rectangle. ■ Write the independent variable (IV) across the top of the rectangle. ■ Divide the rectangle into labeled columns to represent the different levels of the independent variable. ■ Indicate the number of trials in each column. ■ Write the dependent variables (DV) and constants (C) beneath the rectangle.
4. Procedure	List the steps followed to complete the investigation. Check the list carefully for accuracy, completeness, and precision.
5. Results	Complete a data table and an appropriate graph for the data using the following guidelines.
Data Table	■ Make a table containing vertical columns for the independent variable, dependent variable, and derived quantity. ■ Subdivide the column for the dependent variable to reflect the number of trials. ■ Order the values of the independent variable—preferably from the smallest to the largest. ■ Record values of the dependent variable. ■ Compute the derived quantity.
Graph	■ Draw and label the X and Y axes of the graph. ■ Write data pairs for the independent and dependent variables. ■ Determine an appropriate scale for the X and Y axes; subdivide the axes. ■ Plot the data pairs on the graph. ■ Summarize the data trends on the graph.
6. Conclusion	Describe the purpose, major findings, an explanation for the findings, and recommendations for further study. Use six questions to guide your writing of the conclusion. ■ What was the purpose of the experiment? ■ What were the major findings? ■ Was the hypothesis supported by the data? ■ How did your findings compare with other researchers or with information in the textbook? ■ What possible explanation can you offer for the findings? ■ What recommendations do you have for further study and for improving the experiment?

Example

Title:

Hypothesis:

IV: Temperature of Water		
Cold	Room	Hot
5 Trials	5 Trials	5 Trials

DV: Time to dissolve (sec)

C: Brand of effervescent tablet
Amount of water (75 ml)
No stirring
Type of cup

PROCEDURE

In formal papers, you must write the experimental procedure in a paragraph form. You may find it easier to begin by listing the materials you used and the steps you followed in conducting the investigation. Later, you can re-write the lists as a paragraph. Strategies for helping you write clear precise procedures are described in Chapter 3. Select the most appropriate format, paragraphs or lists, for your report.

Example:

One Brand X effervescent tablet was placed in a plastic glass containing 75 ml of cold water and the dissolving time recorded in seconds. Five trials were conducted with cold water. Similarly, dissolving times were determined in room temperature and hot water.

RESULTS

Data collected from an experiment are displayed in simple data tables and appropriate graphs. When you construct a data table, you should make columns for the independent variable, dependent variable, and derived quantity (average dissolving time). Repeated trials are shown in subdivisions of the dependent variable column. Depending upon the type of data, continuous or discrete, you may construct a line graph or bar graph to show the data. Graphs can be constructed by following the five stage sequence of

- drawing and labeling the X and Y axes of the graph
- writing data pairs for the independent and dependent variables
- determining appropriate scales for the X and Y axes and subdividing the axes
- plotting the data pairs on the graph, and
- summarizing the data trends

CONCLUSION

Writing a conclusion is easy when you use a series of questions to guide your thinking.

Q1. What was the purpose of the experiment?
The purpose of the experiment was to determine the effect of water temperature on the dissolving time of Brand X effervescent tablets.

Data Table

Temperature of water	Time to dissolve (sec) Trials 1	2	3	4	5	Average time to dissolve (sec)
ice	98	104	107	96	105	102
room	43	35	46	46	30	40
hot	24	27	19	19	27	23

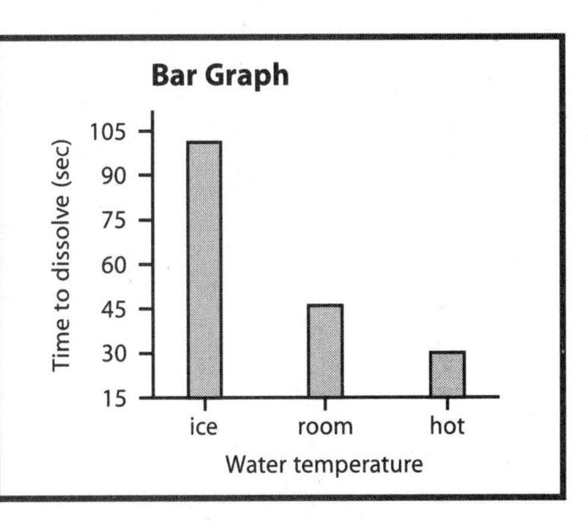

Q2. What were the major findings?
At higher temperatures, Brand X tablets dissolved faster. For each rise in temperature, dissolving time was further reduced.

Q3. Was the research hypothesis supported by the data?
The data supported the research hypothesis that effervescent tablets would dissolve faster in warm water than in cold water.

Q4. How did your findings compare with other researchers or with information in the textbook?
The findings agree with the solubility rule that solids dissolve faster in warm solvents.

Q5. What possible explanations can you offer for the findings?
Because molecules move faster in warm water, they would strike the tablet more frequently and tear it apart more quickly.

Q6. What recommendations do you have for further study and for improving the experiment?
Additional experiments could be conducted to determine the dissolving rate of other brands. The experiment could be improved by insulating the cups to reduce heat exchange with the room.

EVALUATING YOUR REPORT

After you complete each part of the report, such as the title/introduction, you can evaluate the section using the criteria listed in Table 6.2 *Checklist for Evaluating a Simple Report* and make revisions.

Related Web Sites

http://www.dade.k12.fl.us/us1/science/prod03.htm

http://www.isd77.K12.mn.us/resources/cf/SciProjIntro.html

http://www.access.digex.net/~schapman/classes/labs.html

http://www.eduzone.com/Tips/science/SHOWTIP2.HTM (report section)

TABLE 6.2 Checklist for Evaluating a Simple Report

Criteria/Value	Self	Peer/Family	Teacher
Title/Introduction			
Correct title			
Rationale			
Purpose			
Hypothesis			
Experimental Design			
Name/Level/Units of independent variable			
Control			
Repeated trials			
Name/Units of dependent variable			
Procedures			
All steps, equipment, and materials included			
Repetitions for repeated trials and levels of IV			
Spelling/Grammar			
Results—data tables			
Labeled vertical column for independent variable			
Labeled vertical column for dependent variable			
Labeled vertical column for derived quantity			
Correct values of IV, DV, derived quantity			
Results—graphs			
Correct label/Unit/Scale for X axis			
Correct label/Unit/Scale for Y axis			
Data pairs correctly plotted			
Data trends summarized			
Conclusion			
Purpose of experiment			
Major findings			
Support of hypothesis by data			
Comparisons/Explanations			
Recommendations—Further study/improvement			
Spelling/Grammar			

CHAPTER 7

Using Library Resources

Objectives

- Identify critical information for documenting a source and use an approved style format for documentation.
- Identify critical information to record on note cards for different types of information.

  general sources | technical manuals and procedures
  scientific research | interviews with community members
- Complete note cards on a variety of sources.
- Use library resources and questions to generate and refine ideas for an independent research project.
- Use a checklist to evaluate note cards, and to identify needed improvements.

National Standards Connections

- Identify questions and concepts that guide scientific investigations (NSES).

Good scientific research blends the old and the new. First, scientists learn as much as possible about the topic and the experiments conducted by other scientists. Then, they look for ways to expand their understanding, perhaps by investigating a new question or resolving conflicting data. Scientists continually use books and articles from libraries and talk with other experts. In this chapter, you will use information from the library and community experts to answer a variety of questions (see Table 7.1 *Five-Stage Model for Relating Library and Scientific Research*). By blending your interests with the new information discovered in the library, you will design and conduct a better scientific research project.

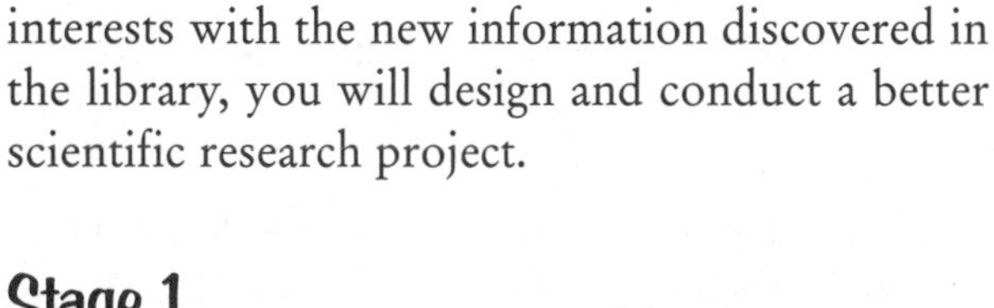

Stage 1

ESTABLISH AN INTEREST

Generally, scientists investigate topics in their areas of expertise and interest. Similarly, you will design and conduct better research in an area of

TABLE 7.1 Five-Stage Model for Relating Library and Scientific Research

Stage	Use appropriate library resources	Use library research skills	Relate library & scientific research
1. Establish an interest	■ Popular magazines ■ Newspapers	■ Documenting ■ Scanning ■ Making note cards	■ What **general topic** (X) interests you? ■ What **general action** of (X) interests you?
2. Narrow the topic	■ Textbooks, K–12 ■ General references: books, encyclopedias, dictionaries, handbooks	■ Using descriptors ■ Locating information in books, card catalogs, general indexes ■ Documenting ■ Scanning ■ Making note cards	■ What **specific topic** (X) interests you? ■ What **materials** are readily available for you to conduct experiments on (X)? ■ What **specific action** of (X) interests you? ■ How could you **measure** or **describe** the action of (X)?
3. Clarify the variables (Optional)	■ Scientific indexes ■ Scientific abstracts ■ Scientific journal articles	■ Using scientific indexes ■ Documenting ■ Scanning ■ Making note cards	■ What **variable** will **you change** to conduct experiments on (X)? ■ What **specific changes** will you make in (X)? ■ What **action** of (X) will you **investigate?** ■ What **specific observations** or **measurements** will you make on (X)?
4. Refine the procedures (Optional)	■ Laboratory manuals, K–12 ■ Handbooks and manuals ■ Sourcebooks ■ Community resources	■ Using community resources ■ Conducting interviews ■ Documenting ■ Scanning ■ Making note cards	■ What specific **materials** or **organisms** will you use? ■ What specific **procedures** will you follow? ■ How will you **collect** and **analyze data?**
5. Interpret the unexpected	■ Appropriate resources	■ Using library skills	■ How can you **explain unexpected** events?

your interest. Before choosing a topic, answer these questions:

1. Where would you like to work on your project—home, school, part-time job, or special research lab?
2. Would you prefer a theoretical project or one that is more practical?
3. What is your favorite school subject?
4. If science is your favorite area, which science do you prefer?
5. What are your hobbies?
6. Have you read a journal article or book that appealed to you? What was it about?
7. Have you seen an interesting television program or film about science? What was the topic?
8. What are your career interests?

Think of your interests and where you plan to work as potential clues of a treasure hunt—in search of "just-the right" research project.

Referencing Information

Did you think of an interesting book, article, or television program you had seen? If so, did you remember the title or the author? Probably not, however, you might remember other things, such as the color of the book or the time of day. Unfortunately, books are not catalogued by color. With all the information available today, it is important to keep reference information, such as titles and author. With this information, you can return to the resource to find additional information. When writing papers, you will also need this information to give credit to the work and ideas of others. To fully credit other people's work, you will need the information summarized in Table 7.2 *Documenting Sources*. The specific way you place the information in a bibliography or footnote will depend upon the **style manual** used. The examples in this book are written using the style of the *American Psychological Association.* Other style manuals include *Modern Language Association, Chicago Manual of Style,* and *Turabian.* Each style manual places the information in a different order. For most school projects, you should record your references using the style that is taught in your school's English classes. Sometimes, however, a particular reference style is required by a science fair or other competitive event. If you plan to enter your experiment in a competition, check with your teacher about special requirements.

Taking Notes

You will find it easier to take and use notes if you use a similar format each time. One approach is to mentally divide your paper, computer screen, or index card into four parts—organizer, reference, information, and reactions (see Figure 7.1).

Organizer (A)

At the top of the page write information needed to arrange the cards or to return to the source later. Useful items include the topic, card number, location, and call number. The **topic** is the major idea or subject described in the resource. The **card number** is the sequence number of the card for the same reference. If you have more than one card for the same reference, this number helps you keep them in order. The **location** is where the resource can be found, such as a school or community library or your science classroom. The **call number** is the number used by the library to catalog and place the books on shelves; the number will help you find a resource if you need it again. Look at the note card on Af-

TABLE 7.2 Documenting Sources

Book:	author, title, place of publication, publisher, edition, copyright date, pages.
Magazine:	author (if available), title of article, name of magazine, date, pages.
Newspaper:	author (if available), title of article, name of paper, date, pages.
Film/Media:	producer, director, title, date, place of publication, publisher, type of media (film, television, computer program).
Unpublished manuscript:	author, title, date; if paper was presented at a meeting add the name, location, and date of the meeting.
Journal article:	author, title of article, journal title, volume and issue number, date, pages.
Abstract:	information needed on the original article is author, title of article, journal title, volume and issue number, date, pages; information needed on the compilation is journal title, volume and issue number, date, pages, abstract number.
Personal Interview:	name of person, topic, location, and date.

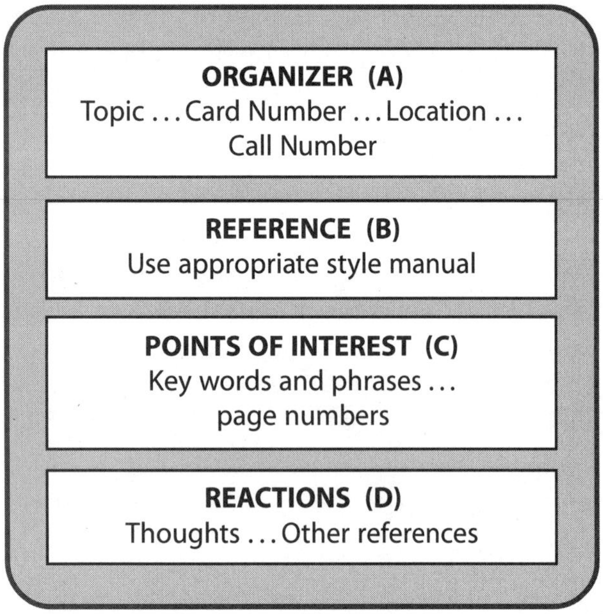

ORGANIZER (A) Topic . . . Card Number . . . Location . . . Call Number
REFERENCE (B) Use appropriate style manual
POINTS OF INTEREST (C) Key words and phrases . . . page numbers
REACTIONS (D) Thoughts . . . Other references

Figure 7.1 General Format for Note Card.

rican violets found in Figure 7.2. What is the call number of the book? Where is it located?

Reference (B)

Next, record the information needed to document the source, such as the author, title, and publisher. For specifics about documentation, use the style manual recommended by your teacher. Remember, different style manuals require different information and record it in different patterns. In Figure 7.2, the reference follows the style of the *American Psychological Association.* Who is the author of the book? What is the title? When and where was it published?

Points of Interest (C)

Use key words and phrases to record the major points and ideas you found interesting. Try to write in your own words and avoid copying. If the author's wording is so important that you need to copy it, be sure to put quotation marks around the quoted words. Make the cards neatly so that you can read them several weeks or months later. Be sure to record page numbers; these are needed to accurately document information and to return to sources. In Figure 7.2, what is the first point made about African violets? Which of the points did you find interesting?

Reactions (D)

Use a few adjectives to describe your reaction to the article. If you found it interesting, state why. If the author gave **other references** that interested you, list them. Later, you may want to find the sources and read them. In Figure 7.2 why was the note taker interested in African violets? Were there other references you might consult?

Using the Library to Establish Your Interest

Browse your science classroom, library, or home for interesting reading. Look for copies of popular magazines, professional teaching magazines, or research papers of other students.

Among the popular magazines readily available in school or community libraries are *Smithsonian, Discover, Nature, Psychology Today, Popular Mechanics, Popular Photography, Popular Science,* and *Field and Stream.* You will find recent copies of these publications in the magazine section. Generally, you can obtain back copies of popular magazines from the librarian. Many of your neighbors or relatives may also have copies.

The school library, science department, or your teacher may subscribe to professional teaching magazines such as *The Science Teacher, Science Scope, The American Biology Teacher,* and *The Journal of Chemical Education.* These publications contain useful news briefs and descriptions of experiments. Ask your teacher where you can find copies.

Don't overlook copies of science research projects done previously by students at your school. Approach these 'amateur publications' just as a professional scientist approaches published journal articles. Look for unanswered questions, conflicting data, and interesting ideas you might investigate.

When you find interesting articles, make a note card using the format illustrated in Figures 7.1 and 7.2. After you have made several cards,

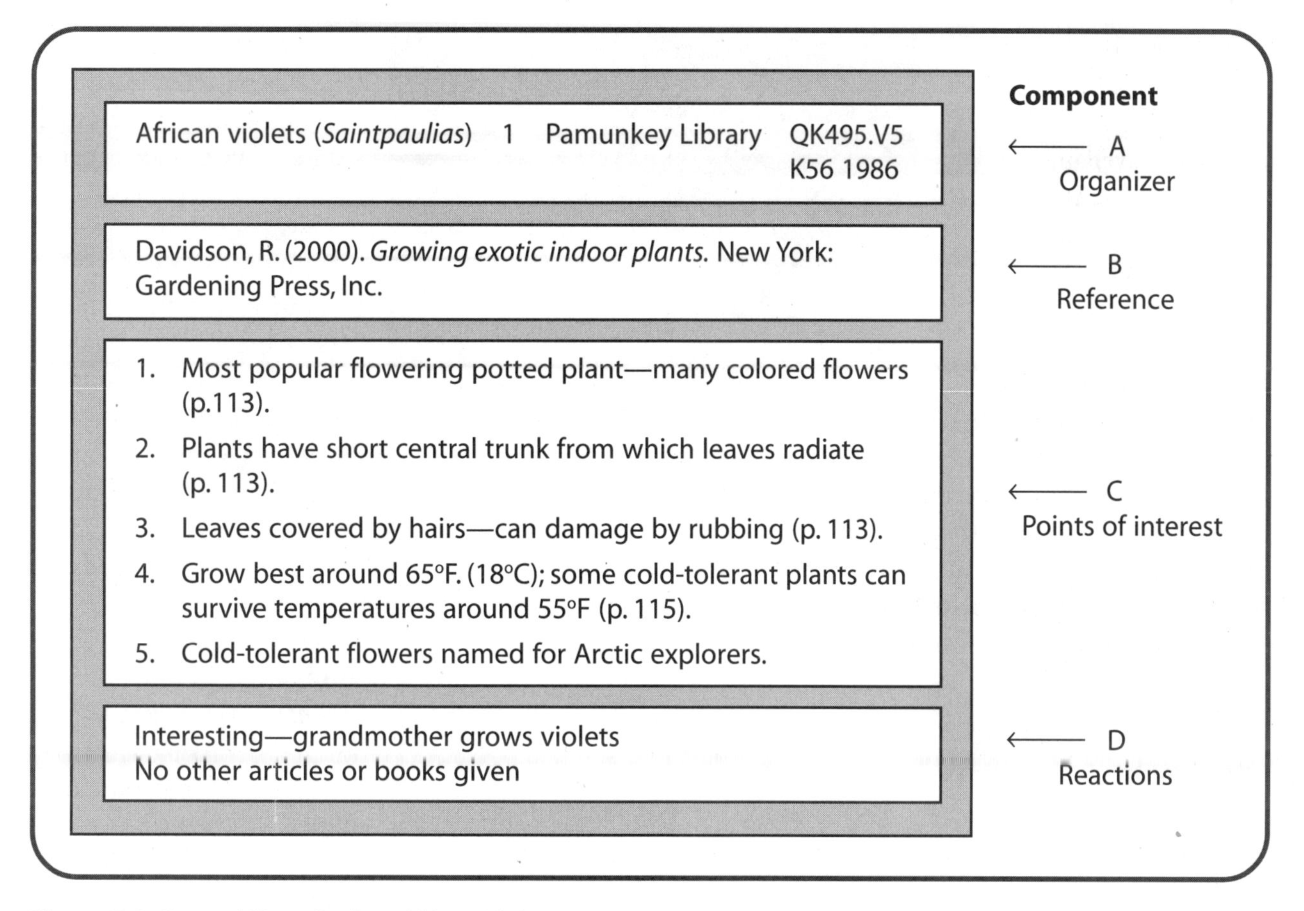

Figure 7.2 General Note Card on African Violets.

reflect upon what you have learned. Respond to the following questions.

- What general topic (X) interests you?
- What general action of (X) interests you?

Compare your responses to the questions with the examples below. How are they similar? different?

General topic	General action
Plants	Growth
Disinfectants	Kills germs
Music	Driving accidents
Glue	Sticking wood
Tobacco	Cellular development
Cold weather	Effectiveness of trumpet oil
Planaria	Regeneration

Stage 2

NARROW THE TOPIC

After you have established an interest, such as the growth of plants, you will need to narrow the topic. After all, there are an unlimited number of experiments you could do about the growth of plants, and you need only one!

Science textbooks are a great place to look for general information. You can use your own, or perhaps those of older or younger friends who are studying a different subject. Think of a biology, chemistry, physics, or earth science textbook as a mini-encyclopedia of knowledge. Your teacher may have shelves of textbooks in the classroom, or there may be a collection in the science department. Often, college professors

have sample books which you might use.

Although textbooks are readily available, finding the information you need is more challenging. You will have to use clues found in the table of contents, glossary, and index. Suppose you needed more information about growth of plants. You might not find the exact words, growth of plants, in either the table of contents or the index of a science textbook. However, you can often find other words that may help you find information about plant growth. For example, you might find the following words or phrases.

© 1998 PhotoDisc, Inc.

Table of Contents	*Index*
Requirements of plants & animals (Chapter 1)	Breeding for better traits (pp. 244–45)
Structure of plants (Chapter 8)	Growth (p. 87)
Increasing the Plant Population (Chapter 11)	Flowering (pp. 307–13, 380)
	Growing—above ground stem (p. 103)
	Growing—underground stem (p. 105)
	How to grow from seeds (p. 18)
	Nutrients for growth (p. 267)
	Growth with grafting (pp. 213–15)

Use the table of contents and index to find the information. Quickly scan the material. If you find a good section, read it carefully and take notes using the note card format illustrated in Figures 7.1 and 7.2. Remember, you do not need to copy every word. Write only the most important ideas, such as those found in the topic sentences. Review several textbooks and make note cards on general information related to your topic.

ORGANIZING IMPORTANT INFORMATION

As you review textbooks, you'll probably encounter many new words. It's important to have a way to organize the words so that you can see how they are related. Then, you can decide which words might be useful in helping you find more information about your topic of interest. One way to organize these new words is to divide your paper into two columns. In the left column, put your **topic** of interest such as *plants.* In this right column, put the **action** which interests you such as *growth.* Under each of these columns put the new words or terms you have found.

Topic *(Plant)*	**Action** *(Growth)*
Leaves	Asexual reproduction
Flowers	Sexual reproduction
Roots	Spores
Stems	Seeds
	Pollen
	Cuttings

Making lists is only one way to organize information. Another way is to make a concept map or a drawing of how the concepts fit together. Concept maps are basically circles and lines, so begin the map by drawing two circles in the middle of the page. Label the circle on the left **topic** and the circle on the right **action.** Write the specific topic (*plants*) that interest you inside the left circle and the specific action (*growth*) that interests you in the right circle. As you find important words, add them to the concept map. For example, a concept map using the words listed above, might look like the one in Figure 7.3. How is the concept map similar to the list of terms? How is it different? What advantages do you see for making concept maps?

One advantage of a concept map is that it shows how the terms are related. From a list you

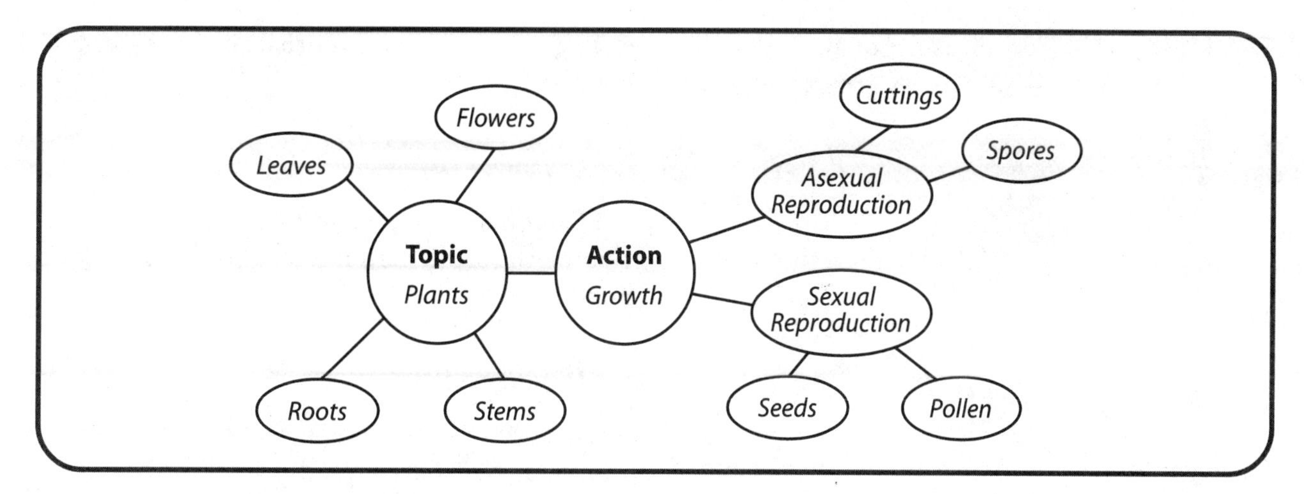

Figure 7.3 Concept Map on Plant Growth.

cannot determine specific examples of asexual and sexual reproduction of plants. With a concept map, however, you can easily see that cuttings and spores are examples of asexual reproduction and that seeds and pollen are related to sexual reproduction.

Once you have made a list or concept map of the major new terms you have learned, it is time to visit your school or community library. Although these libraries will not have textbooks, they will have other valuable sources of information, such as general and scientific encyclopedias, handbooks, dictionaries, and a variety of magazine articles.

Part of a librarian's job is to help people find the books and magazines they need. Here is where your list or concept map of important terms will be very valuable. You can ask the librarian a general question, such as "What reference books do you have on (plants)?" or "How can I learn more about (plant growth)?" You can also use the important terms you've found as clues or descriptors for card catalog systems and for indexes to magazines and journals. The *Reader's Guide to Periodical Literature* is your best single source for finding information in general or non-technical magazines. It can quickly tell you what articles on particular subjects have been published. If you know an author's name, it can also tell you what else the author has written. The *Abridged Reader's Guide*, for the most common magazines, is often found in schools. Today, these reference guides are available on computers, as well as in print form. Through many on-line and CD-ROM computer data bases you can access information quickly.

Using the Library to Narrow Your Interest

As you find useful information continue to make note cards and add new terms to your list or concept map. Once you have read several new references, think about your general interest, such as the growth of plants. This time, however, narrow your interest by responding to four questions.

Q1. What specific topic about (X → plants) interests you?
Leaf cuttings

Q2. What materials are readily available for you to conduct an experiment on (X → plants → leaf cuttings)?
Different varieties of plants, ages of leaves, rooting solutions, light, fertilizer, hormones, soil

Q3. What specific action of (X → plants → growth) interests you?
Root propagation

Q4. How could you measure or describe the action of (X → plants → growth → root propagation)?
Time for roots to appear, number of roots, length of roots, internal development, amount of vascular tissue

Stage 3

CLARIFY VARIABLES

Once you've narrowed your topic, you can begin the search for experiments conducted by other scientists on the same or similar topics. Scientists report their research through articles published in journals. Because there are hundreds of journals in which research is reported, you will need to use specialized indexes to locate information. These include *General Science Index, Biological and Agricultural Index, Applied Science and Technology Index, Chemical Abstracts, Physics Abstracts*, and *Biological Abstracts.* Generally, these specialized indexes and the journals they reference will not be available in your school library. However, you may be able to obtain titles, abstracts, and even complete copies of articles through a variety of CD-ROM and on-line computer services. Also, your school may have special arrangements with local college or university libraries. Ask your teacher or librarian about such arrangements. Always honor the rules that may apply to your use of these college and university libraries, because abuse of this privilege may result in all your fellow students losing access to these valuable library resources.

Referencing and Taking Notes on Journal Abstracts

When making note cards about general information, you should mentally divide your paper or computer screen into four parts—organizer, reference, information, and reactions. This same strategy works well for journal abstracts (see Figure 7.4).

Organizer (A)

Continue to record the same information including the topic, card number, location, and call number.

Reference (B)

Record the information needed to find the resource again. This includes both information about the original journal article and abstracts:

For the original article: author, title of article, journal title, volume and issue number, date, pages;
For abstracts: journal title, volume and issue number, date, pages, abstract number.

Look at the sample note card in Figure 7.4. Who is the author of the complete journal article? What is the title of the complete article? Where was the complete article published? What is the source of the abstract? On which page can you locate the abstract?

Remember, in this book, reference documentation follows the style of the *American Psychological Association.* Be sure you document references using the style manual recommended by your teacher.

Points of Interest (C)

Generally, an abstract is very short and includes brief statements about the purpose of the investigation, the variables investigated, the methods and procedures, and the major findings or conclusion. These parts will not be labeled; you will need to locate them in the abstract. Because of an abstract's brevity, it is often easier to "cut and paste" a copy of the abstract on the note card than to rewrite the components. In Figure 7.4, look at the "cut-and-paste" abstract. What is the independent variable? What is the dependent variable? What was the experimenter's conclusion?

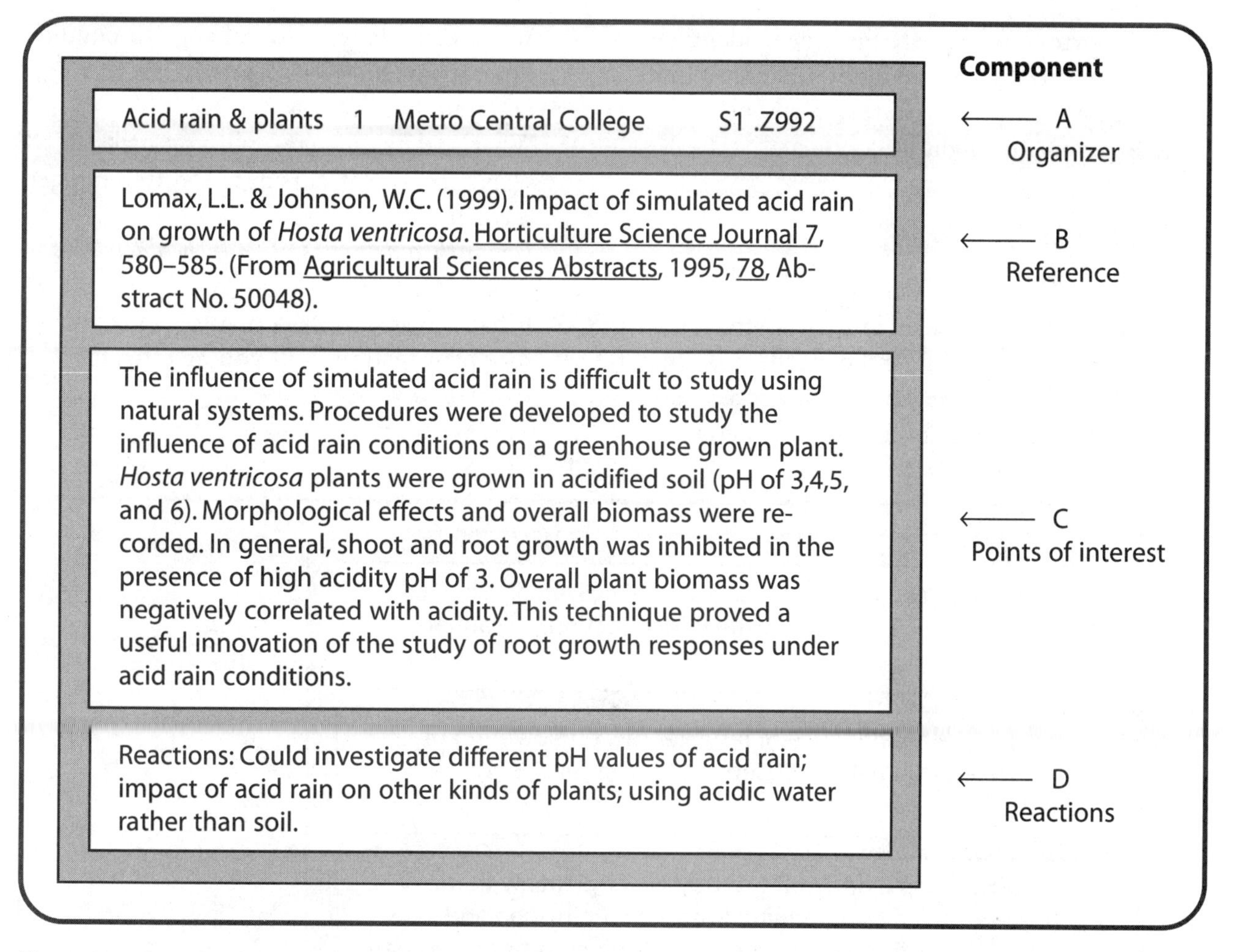

Figure 7.4 Sample Note Card on Journal Abstract.

Reactions (D)

Scientists use journal articles to learn about experiments conducted by others and to look for ways they could expand the knowledge of the subject, perhaps by investigating a new question or resolving conflicting data. Similarly, make notes about experiments you might conduct. In an abstract, you will not find references to other journal articles. From reading the abstract in Figure 7.4, what experiments might you conduct on acid rain?

Referencing and Taking Notes on Journal Articles

Compared with journal abstracts, a complete journal article will be very long. Your challenge will be to record only the most important information. For the **points of interest,** focus upon the following:

- **Purpose of the investigation:** major reason given for doing the experiment and the hypothesis to be tested;
- **Design of the experiment:** the independent variable, levels of the independent variable, dependent variables, control group, constants, repeated trials;
- **Methods and materials:** unusual items used to conduct the experiment and the general procedure (you do not need to write every step);
- **Major findings and conclusion:** what the experimenter learned and potential applications.

Read the sample note card in Figure 7.5. What is the title of the article? In which journal was it reported? What were the independent and dependent variables in the experiment? Which fertilizer mix resulted in greatest flowering? What other experiments might be conducted?

Using the Library to Define Your Variables

After you have read several journal abstracts and articles, focus upon defining the specific variables you will investigate. Use these questions as a tool, substituting your own special area of interest.

Q1. What variable will you change to conduct experiments on (X → plants → leaf cutting)?
Different aged leaves

Q2. What specific changes will you make in (X → plants → leaf cuttings → different aged leaves)?
1-month-old leaves, 2-month-old-leaves, 3 month-old leaves

Q3. What action of (X → plants → growth → root propagation) will you investigate?
Growth of roots

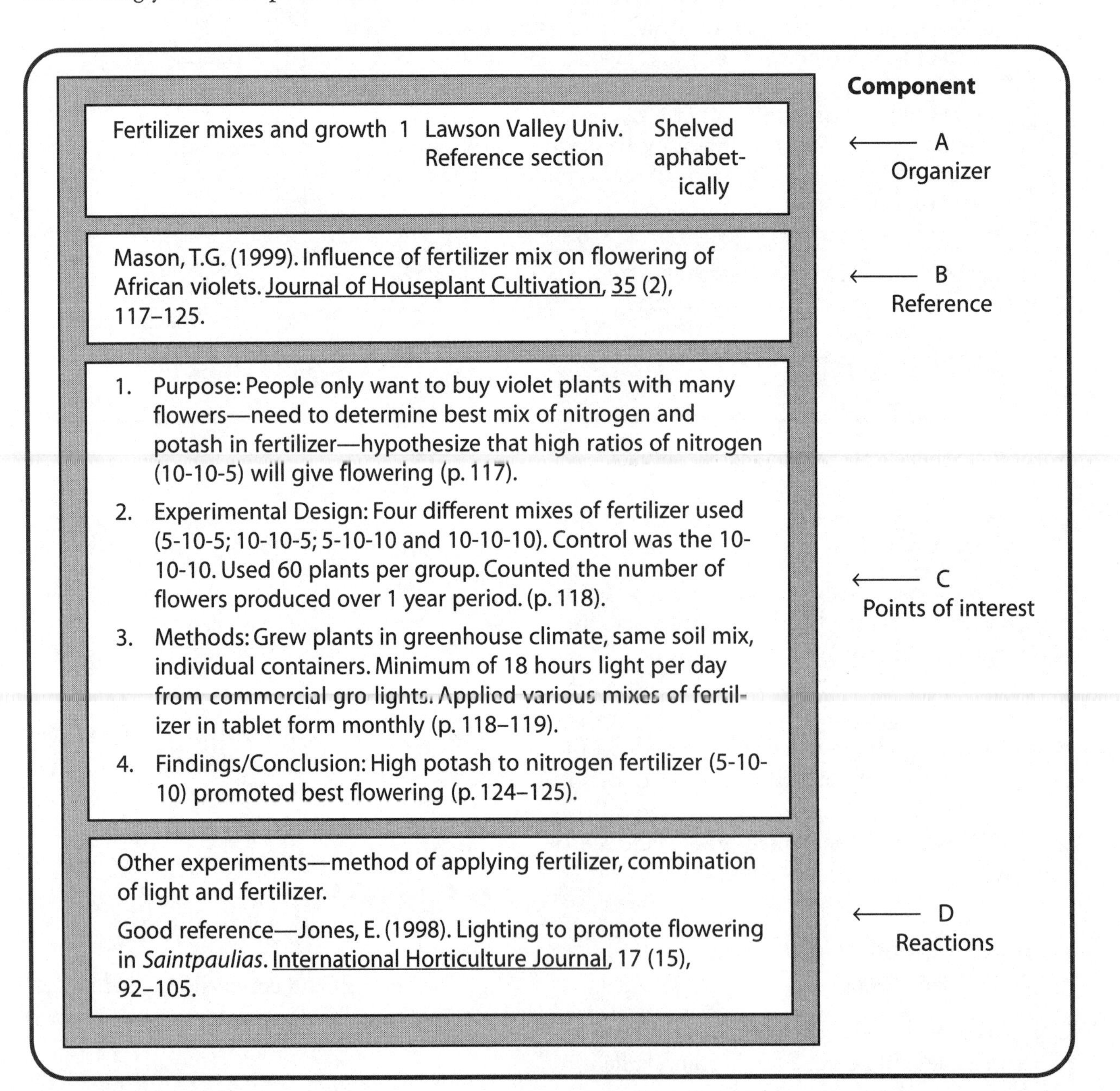

Figure 7.5 Note Card on Journal Article.

Q4. What specific observations or measurements will you make on (X → plants → growth → root propagation → growth of roots)?
Observe the internal development of xylem and phloem
Count the number of roots that emerge

After you have finished, congratulate yourself for identifying a specific independent variable to be manipulated, such as age of leaves, and specific dependent variables to be measured or described, such as the number of roots and the development of xylem and phloem.

Stage 4

REFINE THE PROCEDURES

Now that you have decided upon the variables, you will need to develop procedures for conducting the experiment. Many decisions will need to be made including the specific materials or organisms to be used, the specific procedures to be followed in conducting the experiment and the methods by which you will analyze the data. Information can be obtained from written materials and by interviewing members of the community.

Making Note Cards about Written Procedures

Excellent sources of information on materials and procedures include K–12 laboratory manuals, handbooks, manuals in specific scientific disciplines, and source books written for teachers. Initially, just scan procedures quickly to identify those that are appropriate, safe, and realistic. When you find a procedure that contains readily available and safe materials and which you think you could conduct, make brief notes. On the note card be sure to include the organizer and reference information. For points of interest, concentrate on the major materials and equipment needed, their availability, and a brief synopsis of the steps. Do not try to write the entire procedure. All you need is sufficient information to return to the source if you decide to use the technique. Then, you will probably need to photocopy the information because every detail will be important.

Look at the sample note card about procedures for rooting African violets in Figure 7.6. What is the reference for the procedure? What major procedures for rooting are described? Did the person think they could use the procedure?

Interviewing Community Members and Making Note Cards

There are many people in your community who would be delighted to help you with your research project. Talk to people about your project. Even if they can not help you, they may know someone who could provide information. Talk to your teachers and as many of your friends and classmates as you can. Your parents may also know people who might be helpful.

People who may help can be found at universities and in industries. Depending on your topic you might contact zoos, nature centers, or museums. Check the yellow pages for medical offices, hospitals, and businesses related to your topic. Try greenhouses if your topic is African violets, or electricians and engineers if you are studying some aspect of electricity. Most phone directories also have blue pages that list government agencies that might be able to provide information on your project. Do not feel badly if some people cannot talk to you about your project. Some people just will not have the time or interest in your topic.

Before calling an individual or organization for information, get organized. Make a set of note cards. You can use the same general format as for other note cards, with a few minor changes.

Organizer

Record the topic, card number, location (address), and call number (the phone number).

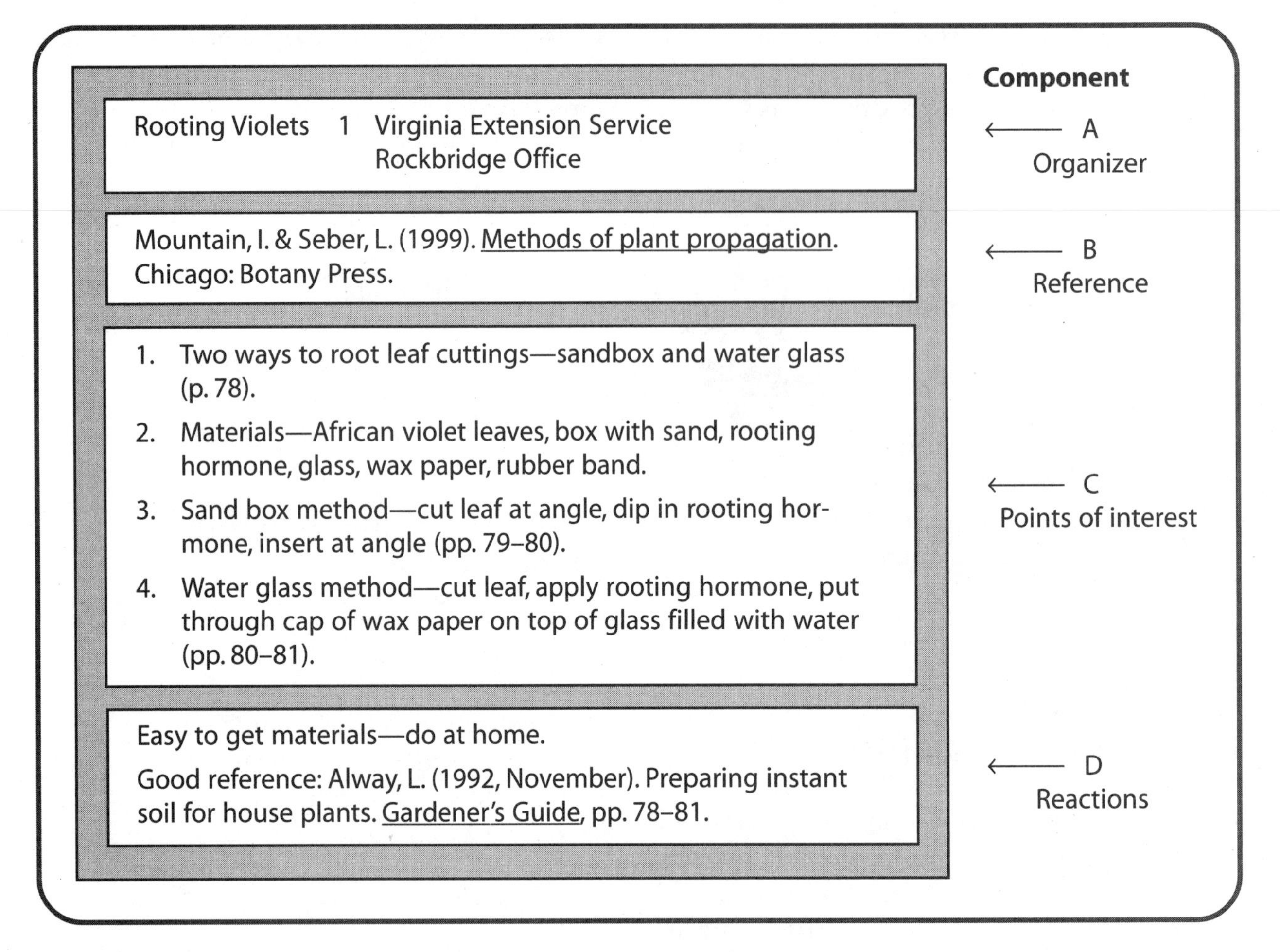

Figure 7.6 Note Card on Procedures.

Reference

Provide reference documentation for the interview including the name of the speaker and the date.

Points of Interest

Make a card for each question you want to ask. Leave space for the person's response.

Reactions

Provide information about the usefulness of the information and other sources mentioned.

Look at the sample note cards used with an interview in Figure 7.7. Who did the person contact? What did they learn about the best light intensity for African violets? about rooting solutions?

When you contact an individual or organization, you represent both yourself and your school. It is important to make a good impression! Here are some tips that will help you make a favorable impression.

- Introduce yourself and provide information about your school and grade.
- Let the individual know what topic you are investigating for a science project and that you would like to ask them a few questions.
- Describe your research problem.
- Ask your questions over the phone or ask for an appointment to discuss your project.
- When taking notes, record important phrases and key words. Tape record the interview if the person agrees.
- Ask the person for suggestions of resources to read or other people to contact for information.

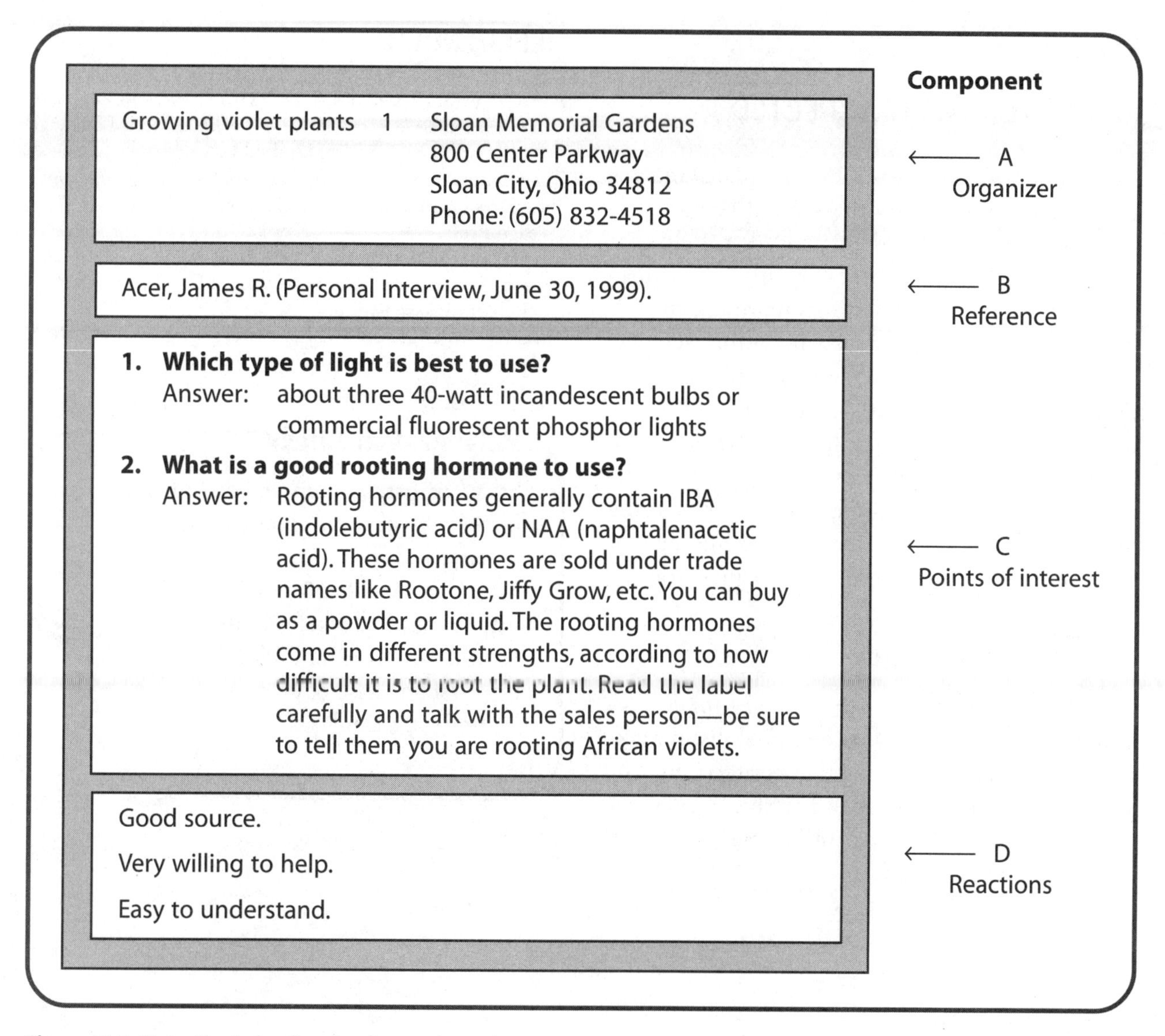

Figure 7.7 Note Cards for Conducting an Interview.

- After your interview, immediately review your notes and add additional information while you still accurately remember details.
- Write a follow-up thank you letter. Most people who took the time to talk to you are interested in what you are doing. Keep them informed and invite them to view your finished project.

Using the Library to Refine Your Procedures

After reviewing the written materials and completing several interviews, respond to the following questions:

- What specific materials or organisms will you use?
- What specific procedures will you follow?
- How will you collect and analyze data?

If you can respond to each question, you are ready to begin experimentation. If not, see which information is missing and talk with your teacher or another adult about where you might locate needed information.

Stage 5

INTERPRET THE UNEXPECTED

Professional scientists realize that plans and results are not necessarily the same. Even so, it is hard to understand when procedures do not work and the experimental results are unexpected. Such unexpected events represent the nature of all human endeavors, including science, and are part of the learning process. Whenever unexpected events occur, continue to use the library and interviewing skills you have learned. Libraries and community organizations are valuable resources for interpreting events, modifying procedures, offering explanations for findings, or proposing options for further experimentation.

Evaluating Note Cards

Use the criteria in Table 7.3 *Checklist for Evaluating Note Cards* to evaluate your skills in making note cards. When using these criteria, use the section that is appropriate for the library resource—general source, scientific research article, procedure, or interview.

REFERENCES

Achtert, W. S., & Gibaldi, J. (1998). *The modern language association style manual.* New York: Modern Language Association of America. (2nd ed.)

The Chicago manual of style: For authors, editors, and copywriters (14th ed.). (1993). Chicago: University of Chicago Press.

Publication manual of the American Psychological Association (4th ed.). (1994). Hyattsville, MD: American Psychological Association.

Turabian, K. L. *A manual for writers of term papers* (6th ed.). (1996). Chicago: University of Chicago Press.

Related Web Sites

http://webster.commnet.edu/mla.htm
http://webster.commnet.edu/apa/apa_index.htm
http://155.43.225.30/workbook.htm
http://sunsite.berkeley.edu/LibraryLand?chil/skill.htm
http://scout.wisc.edu/scout/indextxt.html

TABLE 7.3 Checklist for Evaluating Note Cards

Criteria	Self	Peer/Family	Teacher
Organizational information			
Topic			
Card number			
Location/address			
Call number/phone number			
Reference			
All information included			
Correct style			
Points of interest			
Accurate information			
Paraphrased words/Phrases			
Quotations around author's words			
Page numbers noted			
Appropriate for source			
■ General Source—major points			
■ Scientific Research—purpose, research design, methods, findings, conclusion			
■ Procedures—equipment/materials, availability, summary of steps			
■ Interview—questions, responses			
Reactions			
Thoughts			
Other references, if located			

CHAPTER 8

Analyzing Experimental Data

Objectives

- Distinguish among quantitative, qualitative, ratio, interval, ordinal, and nominal data; give examples of each.
- Select the appropriate measures of central tendency and variation for a given set of data.
- Describe three ways to find the central value of a set of data—mean, median, mode; compute the values for a set of data.
- Describe four ways to report the variation in a set of data range, frequency distribution; compute the values for a set of data.
- Construct an appropriate data table and graph for sets of quantitative and qualitative data.
- Use a checklist to evaluate data tables and graphs and to identify needed improvements.

National Standards Connections

- Use appropriate tools and techniques to gather, analyze, and interpret data (NSES).
- Use mathematics in all aspects of scientific inquiry (NSES).
- Understand and apply measures of central tendency and variability (NCTM).

In sports you speak of practicing the fundamentals. Likewise, there are fundamentals of experimental design. These include being able to identify the basic components of an experiment and to summarize the information in an experimental design diagram. When you analyze the data collected in an experiment, you build upon these fundamentals. Before tackling data analysis, check how well you remember the fundamentals.

REVIEWING BASIC CONCEPTS

Read the scenario in Activity 8.1 about the effect of a chemical on the growth of tomato plants. Identify the independent and dependent variables, constants, control, repeated trials, and hypothesis. Draw an experimental design diagram. Now check your answers with the experimental design diagram in Table 8.1 *Experimental Design Diagram.*

ACTIVITY 8.1 ▪ Analyzing Experimental Data

Scenario

Mary investigated the effect of different concentrations of Chemical X on the growth of tomato plants. Mary hypothesized that if higher concentrations of Chemical X were applied, then the plants would exhibit poorer growth. She grew 4 flats of tomato plants, 10 plants/flat, for 15 days. She then applied Chemical X as follows: Flat A: 0% Chemical X; Flat B: 10% Chemical X; Flat C: 20% Chemical X; and Flat D: 30% Chemical X. The plants received the same amount of sunlight and water each day. At the end of 30 days, Mary recorded the height of the plants (cm), the general health of the plants (healthy/unhealthy), and the quality of the leaves using a four-point scale. Ratings on the leaf quality scale were defined as follows: Rating of 4: Green color, firm, no curled edges; Rating of 3: Yellow-green color, firm, no curled edges; Rating of 2: Yellow color, limp, curled edges; Rating of 1: Brown color, limp, curled leaf.

Raw Data

Height of plants (cm)				Health of plants				Leaf quality			
Concentration of Chemical X				*Concentration of Chemical X*				*Concentration of Chemical X*			
0%	10%	20%	30%	0%	10%	20%	30%	0%	10%	20%	30%
15.0	18.0	12.0	6.0	H	H	H	UN	4	4	2	1
14.0	20.0	10.0	8.0	H	UN	UN	UN	4	3	3	1
13.0	14.0	14.0	5.0	H	H	H	UN	4	3	3	1
15.0	20.0	10.0	4.0	H	H	UN	H	4	4	2	2
15.0	18.0	8.0	4.0	H	H	H	UN	4	4	2	1
17.0	19.0	8.0	5.0	H	UN	UN	UN	4	2	2	1
18.0	18.0	10.0	8.0	H	H	UN	H	4	4	2	2
12.0	18.0	10.0	7.0	H	H	UN	UN	4	4	3	2
19.0	17.0	11.0	8.0	H	H	UN	UN	4	4	2	1
15.0	19.0	12.0	5.0	H	H	H	UN	4	3	2	1

(H = Healthy; UN = Unhealthy)

Data Analysis

Directions

1. Read the scenario of Mary's experiment and identify the independent variable, dependent variables, constants, control, hypothesis, and repeated trials. Draw an experimental design diagram.
2. Classify each of the dependent variables in Mary's experiment as quantitative or qualitative data; justify your answer.

(continued on the following page)

ACTIVITY 8.1 ■ Analyzing Experimental Data *(continued)*

3. Classify each of the dependent variables in Mary's experiment as nominal, ordinal, interval, or ratio data; justify your answer.
4. For each of the dependent variables in Mary's experiment, describe the most appropriate measures of central tendency and variation.
5. Construct a data table for displaying the measures of central tendency, variation, and number for each set of raw data. Compute the appropriate measures of central tendency and variation, and enter them into the table. Construct an appropriate graph.

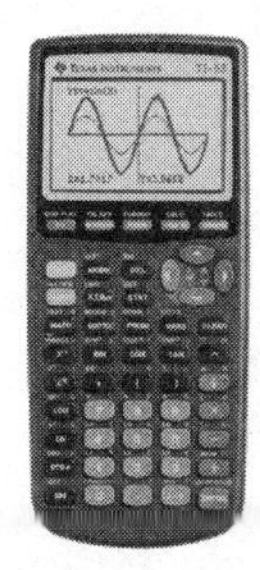

USING TECHNOLOGY

1. In the **STAT** mode of your calculator, enter the values for heights of plants for 0% in List 1, for 10% in List 2, for 20% in List 3, and for 30% in List 4. (See Appendix A, *Using Technology,* for additional help in using the graphing calculator).
2. In the **STAT** mode select CALC (for calculate) and then 1 VAR (for 1-variable statistics). Depending on the brand of your calculator, you will need to enter the desired list number (e.g., L1), or SET the 1-variable x-list to the desired list number before selecting 1 VAR. *Repeat* the selection process for each set of data by changing the list number.
3. Among the calculated values provided are the mean and median for the data set. Calculators that also provide a value for the mode will display the highest mode if there is more than one mode for that set of data. In addition, the range can be calculated from the maximum and minimum values given for each data set.
4. Use appropriate values to construct your summary data table and bar graph.
5. Now, repeat these steps for the leaf quality data. In the **STAT** mode of your calculator, enter the rating values (1, 2, 3, 4) for the various chemical concentrations in Lists 1–4. Sort each list in ascending order and identify the median value for each chemical concentration.
6. Use the appropriate values to construct your summary data table and frequency distribution.

TABLE 8.1 Experimental Design Diagram

Title: The Effect of Various Concentrations of Chemical X on the Growth of Tomato Plants
Hypothesis: If higher concentrations of Chemical X are applied, then tomato plants will exhibit poorer growth.

IV: Concentration of Chemical X			
0% (Control)	10%	20%	30%
10 plants	10 plants	10 plants	10 plants

DV: Height of plants (cm)
Health of plants (healthy/unhealthy)
Leaf quality (scale of 1–4)

C: Amount of sunlight
Amount of water
Amount of pre-experiment growth (15 days)
Length of experiment (30 days)

TYPES OF DATA

In her experiment, Mary recorded three different dependent variables—the height of the plants, the health of the plants (healthy/unhealthy), and the leaf quality (scale of 1–4). Each of the dependent variables can be classified as quantitative or qualitative data.

Quantitative Data

Quantitative data are represented by a number and a unit of measurement that is based upon a standard scale with equal intervals, such as the Metric system of measurement or the Arabic system of numbers. Examples of quantitative variables are the height of a person in meters, the mass of rabbits in kilograms, and the number of seeds that germinated. Quantitative variables may be continuous or discrete. **Continuous quantitative data** are collected using standard measurement scales that are divisible into partial units, for example, distance in kilometers and volume in liters. **Discrete quantitative data** are collected using standard scales in which only whole integers are used, for example, the number of wolves born in a given year or the number of people that can touch their toes. In this chapter, the same statistical techniques will be used with continuous and discrete quantitative data. As shown in the examples on the next page, and explained later, graphical presentations differ.

Quantitative data can be further subdivided based on the zero point of the measuring scale. When quantitative data are collected using a standard scale with equal divisible intervals and an absolute zero, it is called **ratio data**. Examples include the temperature of a gas on the Kelvin scale, the velocity of an object in m/sec, and distance from a point in meters. If the scale does not have an absolute zero, the data are called **interval data.** A common example is the temperature of a substance on the Celsius scale. On this scale, changes in water temperature from 90 to 95 degrees and from 60 to 65 degrees represent the same amount of increase in temperature or kinetic energy of the molecules; however, there is no absolute zero because water molecules are still moving at 0°C. In fact, a substance must reach –273°C (0 Kelvin) before molecular motion ceases. In this chapter, the same statistical and graphical techniques are used with ratio and interval data. Mathematically, however, ratio data can be used in a ratio and proportion; whereas, interval data cannot. This is why you convert the temperature of gases in degrees Celsius to Kelvin before solving problems with Charles Law or the Perfect Gas Law.

Continuous Quantitative Data

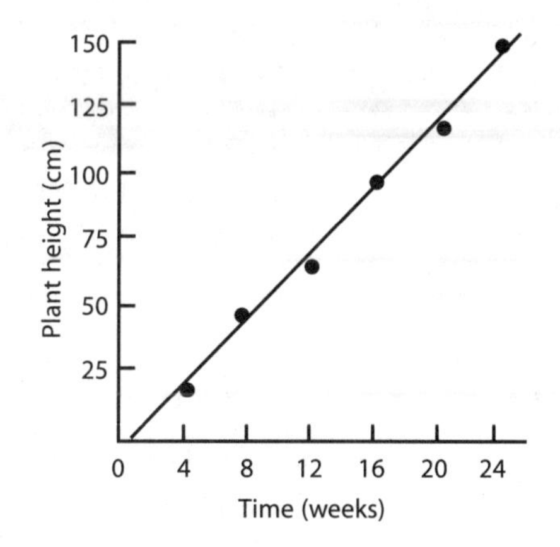

Discrete Quantitative Data

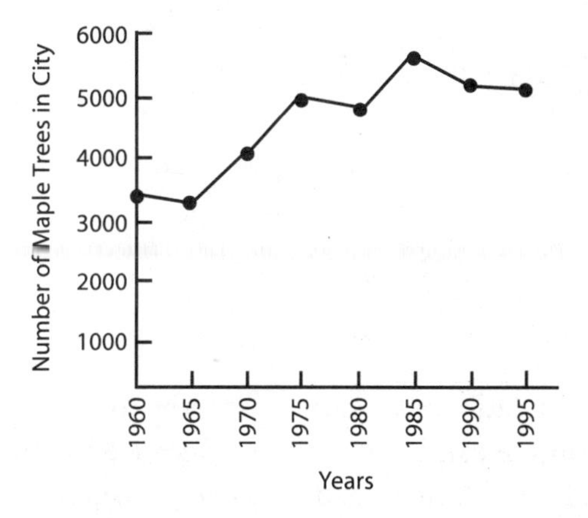

Qualitative Data

Qualitative data are collected by classifying into categories; the categories may be discrete categories represented by a word or "number" label or measurements made with a nonstandard scale with unequal intervals. Examples include the gender of an organism and the color of an individual's eyes. The discrete categories are defined by the experimenter and may be based on a literature review or reflect a synthesis of many observations made during experimentation.

Subdivisions of qualitative data are based on the ability to rank order the categories. **Nominal data** exists when objects have been named or placed into discrete categories that cannot be rank ordered, for example, gender (male/female) and the color of hair (red, black, brown). **Ordinal data** exists when objects are placed into categories that can be rank ordered. For example, the activity of an animal could be rated on a scale of 1 to 5, with 5 representing a very active animal. Another example of ordinal data is Moh's Hardness Scale for Minerals. Different statistical techniques and graphical presentations are used for nominal and ordinal data.

Classifying Data

Using the information in Activity 8.1, classify the three variables in Mary's experiment as quantitative or qualitative data, identify the level of measurements of each variable, and justify your answers.

DESCRIBING DATA

Statisticians describe a set of data in two general ways. First, they compute a **measure of central tendency** or the one value that is most typical of the entire set of data. Second, they describe the **variation**, or spread within the data. Because the types of scales used to collect quantitative and qualitative data differ, the appropriate measures of central tendency and variation also differ (see Table 8.2 *General Overview of Data Analysis*).

Measures of Central Tendency

Three different measures of central tendency are available—the mean, median, and mode. The level of measurement of the data determines which measure is appropriate. Definitions and proce-

TABLE 8.2 General Overview of Data Analysis

Type of information	Quantitative data Interval or Ratio	Qualitative data Nominal	Qualitative data Ordinal
What is the most typical or central value?	Mean	Mode	Median
What is the variation or spread?	Range	Frequency distribution	

dures for calculating the mean, median, and mode are given next.

Mode

The value of the variable that occurs most often. It is used for data at the nominal, ordinal, interval, or ratio levels. If two or more values tie in having the most cases, report both modes.

Examples:

7	15	10
6	13	10
5	12	8
5	11	7
4	9	7
3	9	5
Mode = 5	Mode = 9	4
		Modes = 10 and 7

Median

The middle value, after all of the cases have been rank ordered from highest to lowest. Half of the cases fall above the median value, half below. The median can be used with ordinal, interval, or ratio data but not with nominal data.

Examples:

7	15	10
6	13	10
5	12	8
5 ☜	11 ☜	7 ☜
4	9	7
3	9	5
Median = 5	Median = 11.5	4
		Median = 7

Mean

The arithmetic average or the sum of the individual values divided by the number of cases. The mean can only be calculated for interval or ratio data.

$$\text{Mean} = \frac{7 + 6 + 5 + 5 + 4 + 3}{6} = \frac{30}{6} = 5$$

$$\text{Mean} = \frac{15 + 13 + 12 + 11 + 9 + 9}{6} = \frac{69}{6} = 11.5$$

$$\text{Mean} = \frac{10 + 10 + 8 + 7 + 7 + 5 + 4}{7} = \frac{51}{7} = 7.3$$

For ratio and interval data, the mean, median, and mode can be calculated. Because the mean is usually the best measure of central tendency, it is generally reported for ratio and interval data. The exceptions are those sets of data in which there are a few extreme values that would distort the mean; then the median or mode may be a more accurate measure of central tendency. For ordinal data, both a median and a mode can be calculated. However, the median is generally reported as the better statistic. The mode is the only appropriate measure of central tendency for nominal data.

Measures of Variation

Because you frequently compute averages, you are probably more familiar with the concept of central tendency than the concept of variation within data. Simple measures of variation are the range for a set of quantitative data and the frequency distribution for a set of qualitative data. The **range** is computed by finding the difference between the smallest (minimum) and the largest (maximum) measures of the dependent variable, for example, plant height. Although this value is easy to calculate, being able to interpret it is more difficult. For example, two experimental groups may have equivalent means yet be very different. Consider John's data on 25 tomato plants grown with a red ground cover and 25 tomato plants grown without a ground cover.

	Red ground cover
Mean Height	15.0 cm
Range in height	10.0 cm
Maximum (largest plant)	18.0 cm
Minimum (smallest plant)	8.0 cm
Number	25 plants
	No ground cover
Mean Height	14.9 cm
Range in height	2.0 cm
Maximum (largest plant)	16.0 cm
Minimum (smallest plant)	14.0 cm
Number	25 plants

Although the means or average heights of the two plant groups were equivalent, the plants grown with the red ground cover exhibited much greater variation.

Frequency Distribution

Similarly, the variation within qualitative data is easier to calculate than to conceptualize. The variation is described through a **frequency distribution** that shows the number of cases falling into each category of the variable, for example, the color of tomatoes produced with different ground covers.

	Red ground cover	
Mode	Pink tomatoes	
Frequency Distribution	Red:	0
	Pink:	12
	Yellow:	8
	Green:	5
Number	25 plants	
	No ground cover	
Mode	Red tomatoes	
Frequency Distribution	Red:	20
	Pink:	5
	Yellow:	0
	Green:	0
Number	25 plants	

Notice how the distribution of colors in the two groups differ. Tomatoes in the red ground cover group are generally pink or yellow, whereas those in the no ground cover group are mostly red. You may also calculate more powerful measures of variation for quantitative data—the variance and standard deviation. Procedures for calculating these measures are described at the end of this chapter and in Chapters 10 and 11.

Table 8.2 summarizes the statistical concepts used to describe measures of central tendency and variation in sets of quantitative and qualitative data. Use the table to determine the most appropriate measures of central tendency and variation to calculate for each of the dependent variables in Mary's experiment about the effect of Chemical X on the growth of tomato plants.

Height of plants
Mean
Range

Health of plants
Mode
Frequency distribution

Leaf quality
Median
Frequency distribution

TABLE 8.3 General Data Table for Descriptive Statistics

Title: The Effect of the (IV) on the (DV)			
	Independent variable		
Descriptive information	**Level 1**	**Level 2**	**Level 3**
Central tendency			
Variation		*(Enter computed values)*	
Number (of trials)			

DATA TABLES FOR DESCRIPTIVE STATISTICS

In addition to learning to compute various descriptive statistics, you will need to construct data tables that communicate the statistics. As illustrated above, a data table can be constructed easily by combining the rectangular portion of the experimental design diagram with a section listing the specific measures of central tendency, variation, and the number of repeated trials (see Table 8.3 *General Data Table for Descriptive Statistics*). A title that communicates the specific variables being investigated should accompany each table.

DATA TABLE FOR PLANT HEIGHT

Using the raw data on plant height included in Activity 8.1, *Analyzing Experimental Data,* calculate the mean, range, and number for each concentration of Chemical X. Construct an appropriate data table and enter the computed values (see Table 8.4 *The Effect of Various Concentrations of Chemical X on the Height of Tomato Plants*).

TABLE 8.4 The Effect of Various Concentrations of Chemical X on the Height of Tomato Plants

	Concentration of Chemical X (%)			
Descriptive information	**0**	**10**	**20**	**30**
Mean	15.3	18.1	10.5	6.0
Range	7.0	6.0	6.0	4.0
Maximum	9.0	20.0	14.0	8.0
Minimum	12.0	14.0	8.0	4.0
Number	10	10	10	10

Because both the independent variable, concentration of Chemical X, and the dependent variable, plant height (cm) are quantitative continuous data, either a bar or line graph can be used. Because the levels of the independent variable are few and widely separated, a bar graph was selected, as illustrated in Figure 8.1.

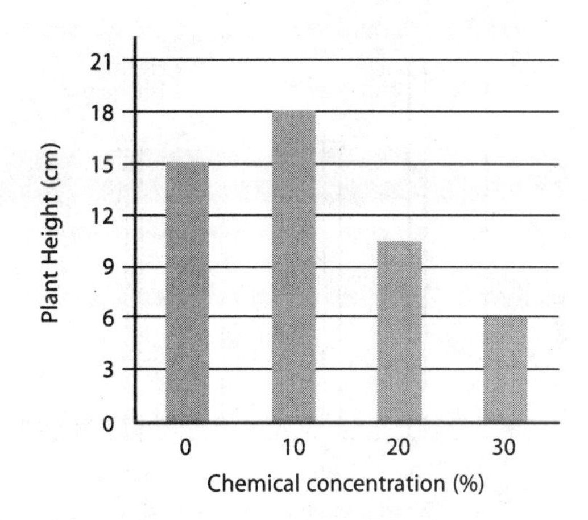

Figure 8.1 Mean Height of Plants Exposed to Various Concentrations of Chemical X.

DATA TABLE FOR PLANT HEALTH

Because the health of the plant was reported as qualitative data at the nominal level of measurement, the mode is the appropriate measure of central tendency. Variation is reported through a frequency distribution. Use the data from Mary's experiment to construct an appropriate data table for summarizing plant health at each concentration of Chemical X (see Table 8.5 *The Effects of Various Concentrations of Chemical X on the Health of Tomato Plants*).

Because the dependent variable is qualitative (nominal), a special type of bar graph, a frequency distribution, is used to graph the data. A frequency distribution shows the number of plants that fall into each category, healthy and unhealthy. On the graph, place the independent variable, concentration of the chemical, on the X axis. Subdivide the Y axis into an appropriate scale for plotting the number of plants that fall into each category of the dependent variable, plant health. Graph the frequency distribution by drawing a vertical bar from a specific concentration of Chemical X to the appropriate number of plants falling into each category, healthy and unhealthy. As illustrated in Figure 8.2, use a key, such as white and shaded bars, to show each category. From the graph, both the frequency distribution and mode for each concentration of Chemical X can be determined.

TABLE 8.5 The Effect of Various Concentrations of Chemical X on the Health of Tomato Plants

	Concentration of Chemical X (%)			
Descriptive information	**0**	**10**	**20**	**30**
Mode	Healthy	Healthy	Unhealthy	Unhealthy
Frequency distribution				
Healthy	10	8	4	2
Unhealthy	0	2	6	8
Number	10	10	10	10

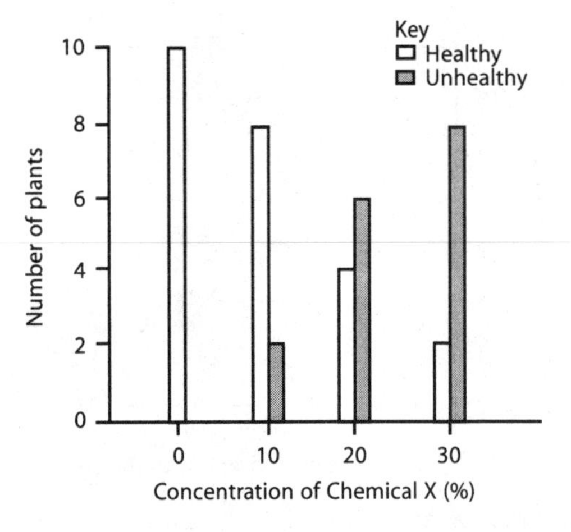

Figure 8.2 Frequency Distribution for Plant Health at Various Concentrations of Chemical X.

DATA TABLE FOR LEAF QUALITY

Mary used a 4-point scale for rating leaf quality. Because the scale involved nonstandard measurements and placement into categories that could be ordered, the median is the appropriate measure of central tendency. Variation is reported through a frequency distribution. Use the raw data on leaf quality from Activity 8.1 to compute the median and frequency distribution (see Table 8.6 *The Effect of Various Concentrations of Chemical X on Leaf Quality*).

Leaf quality represents qualitative (ordinal) data; thus, only a bar graph is appropriate. Again, subdivide the X axis to show the various concentrations of the chemical. Use the 4-point rating scale for leaf quality as subdivisions of the Y axis. Graph the median leaf quality value for each concentration of Chemical X (see Figure 8.3). Using procedures previously described, the frequency distribution for the number of plants falling into each category of leaf quality may also be graphed (see Figure 8.4).

MAKING DECISIONS ABOUT DESCRIPTIVE STATISTICS AND GRAPHS

Now that you have substantially increased your ability to statistically describe your data, how do you determine an appropriate measure of central tendency and dispersion/variation for a specific experiment. How do you decide upon a graphical presentation? Using Table 8.7 *Making Decisions about Descriptive Statistics and Graphs*, ask students to determine if the levels of the independent variable are continuous or discrete. Then, ask them if the dependent variable is quantitative (continuous/discrete), qualitative (ordinal), or qualitative (nominal).

TABLE 8.6 The Effect of Various Concentrations of Chemical X on Leaf Quality

	Concentration of Chemical X (%)			
Descriptive information	**0**	**10**	**20**	**30**
Median	4	4	2	1
Frequency distribution				
Quality 4	10	6	0	0
Quality 3	0	3	3	0
Quality 2	0	1	7	3
Quality 1	0	0	0	7
Number	10	10	10	10

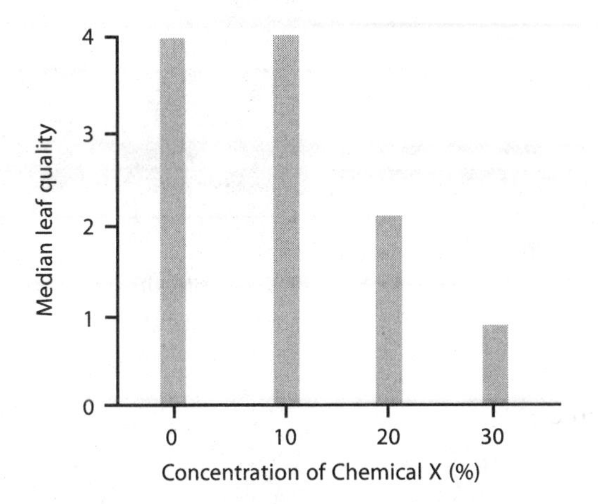

Figure 8.3 Median Leaf Quality for Plants Exposed to Various Concentrations of Chemical X.

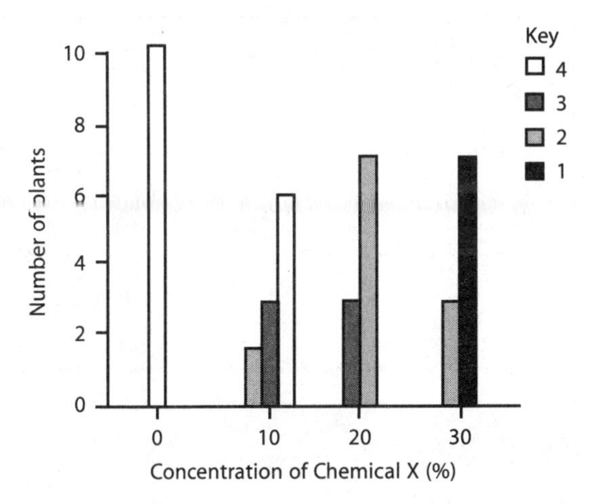

Figure 8.4 Frequency Distribution of Leaf Quality for Plants Exposed to Various Concentrations of Chemical X.

1. *Are the levels of the independent variable continuous or discrete?*
   Continuous means that the levels of the variable are not separate categories and that the intervals between the levels have meaning. An example is the sugar content of a fluid expressed as 0%, 5%, 10%, 15%, 20%, 25%, and 30%. Discrete, however, means that the categories are separate and not continuous. An example of discrete levels of a variable is the sugar content of fluid expressed as low, medium, and high concentration.

2. *Is the dependent variable quantitative (continuous/discrete), qualitative (ordinal), or qualitative (nominal)?*
   If the dependent variable is **quantitative (continuous/discrete)** then three measures of central tendency can be used: the mode, median, or mean. Generally, the mean is used when the data follows a normal distribution (mound shape). When data are skewed, the median may be a more appropriate measure of central tendency. The range is an appropriate measure of dispersion/variation.

   If the independent variable is discrete, then bar graphs are the only option for displaying the data. If the independent variable is continuous, either bar or line graphs may be used. Line graphs, with a line-of-best-fit, are used with continuous quantitative data. Broken line graphs, in which the data points are connected, are used with discrete quantitative data. The same statistical and graphical presentations are used for ratio and interval data.

   If the dependent variable is **qualitative (ordinal),** then two measures of central tendency can be used, e.g. the mode and median. Generally, the median is used. A frequency distribution is used to communicate dispersion/variation within the data. For the graphical presentation, you may make a bar graph of the medians or a frequency distribution.

   If the dependent variable is **qualitative (nominal),** then the measure of central tendency is the mode and the measure of dispersion/variation is the frequency distribution, which is also used for the graphical presentation.

Read Investigation 8.1 *A Sudsy Experience* (see page 88). Then, use the information in Table 8.7 *Making Decisions about Descriptive Statistics and Graphs* to make decisions about appropriate descriptive statistics and graphical displays for the experiment. You should conduct the experiment and display the data through tables and appropriate graphs.

TABLE 8.7 Making Decisions about Descriptive Statistics and Graphs

Levels of Independent Variable	Dependent Variable	Central Tendency	Dispersion/ Variation	Graphical Displays
Continuous	Quantitative (continuous/ discrete)	Mode Median Mean	Range	Bar Graph Line Graph, Line-of-Best Fit (continuous) Line Graph, Broken Line (discrete)
	Qualitative (ordinal)	Mode Median	Frequency distribution	Bar Graph (medians) Frequency distribution
	Qualitative (nominal)	Mode	Frequency distribution	Frequency distribution
Discrete	Quantitative (continuous/ discrete)	Mode Median Mean	Range	Bar Graph
	Qualitative (ordinal)	Mode Median	Frequency distribution	Bar Graph (medians) Frequency distribution
	Qualitative (nominal)	Mode	Frequency distribution	Frequency distribution

Evaluating Your Data Tables and Graphs

In Chapter 5 you learned to use checklists to evaluate your skill in constructing simple data tables and graphs. Similarly, you can use Part One of the checklist in Table 8.8 *Checklist for Evaluating Data Tables with Descriptive Statistics and Accompanying Graphs* to evaluate a data table with descriptive statistics. The checklist can be used with both quantitative and qualitative data. You can also use Part Two of the checklist found in Table 8.8 to determine if you made an appropriate graph to accompany the data table. After you have reviewed your work, ask a friend or family member to give you feedback. Use the feedback to improve your data displays before submitting to your teacher or a competition.

REFERENCES

Landwehr, J.M. & Watkins, A.E. (1994). *Exploring data.* A component of the Quantitative Literacy Series. Palo Alto, CA: Dale Seymour Publications.

Landwehr, J.M., Swift, J., & Watkins, A.E. (1998). *Exploring surveys and information from samples.* A component of the Data-driven Mathematics Series. Palo Alto, CA: Dale Seymour Publications.

McClave, J.T., Dietrich, Frank H. II, & Sincich, Terry. (1997). *Statistics* (7th ed.) Upper Saddle River, NJ: Prentice-Hall, Inc.

Yates, D.S., Moore, O.S., McCabe, G.P. (1999). *The Practice of Statistics: TI-83 Graphing Calculator Enhanced.* New York: W.H. Freeman and Company.

Related Web Site

http://www.mste.uiuc.edu/stat/stat.html

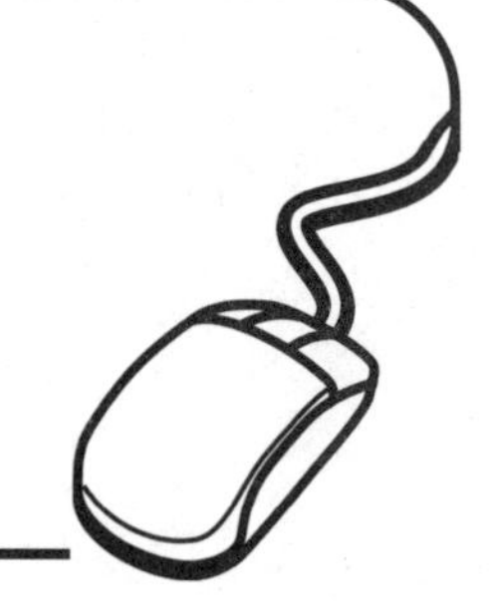

TABLE 8.8 **Checklist for Evaluating Data Tables with Descriptive Statistics and Accompanying Graphs**

Criteria	Self	Peer/ Family	Teacher
Part One—Data Tables			
Title			
Independent variable/Levels/Units			
Units of dependent variable			
Correct descriptive statistics for data			
Central tendency			
Variation			
Number			
Correct calculations of descriptive statistics			
Central tendency			
Variation			
Number			
Part Two—Graphs			
Title			
Correct type of graph for data			
Correct label/Unit/Scale for X axis			
Correct label/Unit/Scale for Y axis			
Data pairs correctly plotted			

INVESTIGATION 8.1 ▪ A Sudsy Experience

Question

How do minerals in water affect its ability to form suds?

Materials for Each Group of Two Students

3	100 ml graduated cylinders
3	pieces of plastic wrap (25 cm^2) torn from roll
3	rubber bands (optional)
50 ml	distilled water
50 ml	water with borax (2 g of borax per 1000 ml of distilled water)
50 ml	water with Epsom salt (2 g of Epsom salt per 1000 ml of water)
1	metric balance
1	bar of Ivory soap
1	device to scrape soap such as plastic knife, metal spoon, or vegetable peeler
1	small container, such as a plate or evaporating dish for scraped soap
1	roll of paper towels or newspapers
1	metric ruler
1	masking tape for labels

Safety

- Wear goggles.
- Wash hands.
- Dispose of chemicals in marked containers.

Procedure

1. Place newspaper or paper towels on work surface in case of spills.
2. Put 50 ml of distilled water in a 100 ml graduated cylinder. Use a piece of masking tape to label the container. Add 1 g of "shaved" Ivory soap.
3. Place the plastic wrap over the top of the cylinder, being careful to seal well. To help insure a tight seal, you may also use a rubber band to secure the plastic wrap. Hold the cylinder straight in front of your body. Place one hand at the top of the cylinder and the other where the plastic wrap ends. Have your partner hold a Metric ruler in front of you; use the length of the ruler as the "length of a shake." Shake the container up and down 10 times.
4. Remove the plastic wrap. Measure the height of the suds column in millimeters (mm). Record the measurement.
5. Place the cylinder in the middle of the work surface. Be sure that you have labeled it so that you will know the type of water inside. After 30 minutes, classify the type of suds formed using the following categories.

(continued on the following page)

INVESTIGATION 8.1 ▪ A Sudsy Experience *(continued)*

LCIN	Light foamy suds, cloudy water, no large soap particles
LCIF	Light foamy suds, cloudy water, few large soap particles
TCIF	Thick suds, cloudy water, few large soap particles
TCM	Thick suds, clear water, many large soap particles

6. Repeat Steps 2 to 5 using 50 ml of water with borax and 50 ml of water with Epsom salt.
7. Enter your data in a class data table. Be sure that at least 10 sets of data are entered. If there are fewer than 10 student groups, then repeat Steps 1–6 to obtain sufficient trials.

Class Data Table I: Height of Suds Column (mm)

	Height of column (mm)													
	Trials (lab work groups)													
Type of water	**1**	**2**	**3**	**4**	**5**	**6**	**7**	**8**	**9**	**10**	**11**	**12**	**13**	**etc.**
Distilled water														
Water with borax														
Water with Epsom salt														

Class Data Table II: Type of Suds Formed (LCIN, LCIF, TCIF, TCM)

	Type of suds formed													
	Trials (lab work groups)													
Type of water	**1**	**2**	**3**	**4**	**5**	**6**	**7**	**8**	**9**	**10**	**11**	**12**	**13**	**etc.**
Distilled water														
Water with borax														
Water with Epsom salt														

Analyzing Your Data

1. Make a summary data table and graph to display the class data on "Height of Suds."
2. Make a summary data table and graph to display the class data on "Type of Suds Formed."

(continued on the following page)

INVESTIGATION 8.1 ▪ A Sudsy Experience *(continued)*

USING TECHNOLOGY

1. In the **STAT** mode of your calculator, enter the values for heights of soap suds for distilled water in List 1, for water with borax in List 2, and for water with Epsom salts in List 3. (See Appendix A, *Using Technology*, for additional help in using the graphing calculator).
2. In the **STAT** mode select CALC (for calculate) and then 1 VAR (for 1-variable statistics). Depending on the brand of your calculator, you will need to enter the desired list number (e.g., L1), or SET the 1-variable x-list to the desired list number before selecting 1 VAR. *Repeat* the selection process for each set of data by changing the list number.
3. Among the calculated values provided are the mean and median for the data set. Calculators that also provide a value for the mode will display the highest mode if there is more than one mode for that set of data. In addition, the range can be calculated from the maximum and minimum values given for each data set.
4. Use appropriate values to construct your summary data table and bar graph.

Extending Your Learning

1. How does Ivory bar soap differ from other brands of soap? Would you get the same results with other types of soap? Why?
2. How did the presence of borax and Epsom salt affect the suds column? Did the data support your hypothesis?
3. What is the chemical formula for Epsom salts?
4. Why is water containing magnesium and calcium ions called "hard water"? How does the presence of these ions affect suds formation? Cleaning?
5. What is the chemical formula for borax? How does the presence of borax affect suds formation? Cleaning?
6. How does a home water softening unit work? What types exist?
7. Why are some types of water softening units not recommended for people with high blood pressure, heart disease, and kidney disease?
8. How could you improve this experiment?
9. What other experiments might you conduct on this topic?

Practice

1. For each of the experimental scenarios below, state the following information.

   A. Type of data (quantitative or qualitative) collected for the dependent variables.
   B. Level of measurement (nominal, ordinal, interval, or ratio) represented by the measures of the dependent variables.
   C. Most appropriate measure of central tendency for describing the dependent variable.
   D. Most appropriate measure of variation for describing the dependent variable.
   E. Construct a data table for the data. Compute the appropriate measures of central tendency, variation, and number; enter them in the table.
   F. Construct an appropriate graph.

The heated soil scenario: Walter placed 1 cup of sand (S), potting soil (P), and a mixture of sand and potting soil (M) into separate pint-size containers. In each of the containers he placed a thermometer so that the bulb was 2.5 cm below the surface. He placed the 3 containers under identical heat lamps for an hour. The original temperature of each jar was 15°C. After heating the jars the first time, the temperatures of the containers were S = 28°C, P = 33°C, and M = 29°C. After heating the jars a second time, the temperatures of the contents were S = 26°C, P = 29°C, M = 29°C. After the third heating, the temperatures were S = 27°C, P = 31°C, M = 22.5°C. Between each heating, the contents of the jars were cooled to 15°C.

The peat moss scenario: Norm wanted to know if adding peat moss to sand would affect its ability to hold water. He put 200 ml of pure sand into container A. He put a mixture of 80% sand and 20% peat moss into container B. Into container C he placed a mixture of 60% sand and 40% peat moss. Finally, he placed a mixture of 40% sand and 60% peat moss into container D. He added water to each container and measured the amount of water the contents would absorb. He dried the sand and peat moss. He repeated the experiment 5 times. He collected the following data.

	Water holding capacity (ml)				
Composition of mixture	**Trial 1**	**Trial 2**	**Trial 3**	**Trial 4**	**Trial 5**
100% sand	74	80	70	71	74
60% sand; 40% peat moss	86	88	90	92	94
40% sand; 60% peat moss	110	116	104	108	112
80% sand; 20% peat moss	84	82	86	82	84

(continued on the following page)

Practice *(continued)*

2. Use the following experimental situation to answer Questions 2A–2D. John raised deer mice. From one litter, he obtained mice with the following masses:

Mouse	Mass (g)
A	1.5
B	2.5
C	3.0
D	2.5
E	2.0
F	1.5
G	2.5

   A. Which type of data is represented (qualitative or quantitative)?
   B. What is the minimum value of the masses of the mice?
   C. What is the range of mouse mass?
   D. Compute the mean, mode, median, and the range of the data.

3. Use the following experimental situation to answer Questions 3A–3C. John sampled the apples on 5 trees in three different orchards, with different varieties, to determine the stage of ripeness. For Variety 1, he counted 70 dark green, 60 yellow-green, 80 pink and 60 red apples. For Variety 2, he counted 40 dark green, 120 yellow-green, 50 pink and 60 red apples. For Variety 3, he counted 10 dark green, 30 yellow-green, 45 pink and 185 red apples.

   A. Which level of measurement does the data represent, e.g. nominal, ordinal, interval?
   B. Construct an appropriate data table.
   C. Construct an appropriate graph.

CHAPTER 9

Communicating Descriptive Statistics

Objectives

- Write appropriate paragraphs of results for quantitative and qualitative data.
- Write an appropriate conclusion for an investigation.
- Use a checklist to evaluate paragraphs of results and a conclusion, and to identify needed improvements.

National Standards Connections

- Recognize and analyze alternative explanations and predictions (NSES).
- Think critically and logically to make the relationships between evidence and explanations (NSES).
- Evaluate arguments that are based on data analysis (NCTM).

Remember the last time you had to write something? Whether that something was background information for an experiment, a sentence about a graph, or a complete laboratory report, what did you think? "I don't know where to start." "What do I need to include?" "I know what it means, but I just can't say it." "What's a conclusion?" These and similar thoughts are common. They reflect the difficulty most people have in moving from unfocused thoughts to precise written language.

In Chapter 8, you learned how to make a data table that included a measure of central tendency and a measure of variation. The terms statisticians use to describe the measures of central tendency and variations in sets of quantitative and qualitative data are summarized in Table 9.1 *Describing Quantitative and Qualitative Data*. If you have forgotten about these types of data analysis, you will find it helpful to review Chapter 8, *Analyzing Experimental Data*.

Before you can write effectively, you must organize your thoughts. There are many ways, such as making an outline, drawing a concept map, or listing the important things you want to communicate. You can also organize your thoughts by discussing your experiment with friends and answering questions they may have. In this chapter, you will use sets of questions as a guide for writing paragraphs about quantitative and qualitative data and for writing a conclusion. These questions are summarized in Table 9.2 *How to Communicate Descriptive Statistics*.

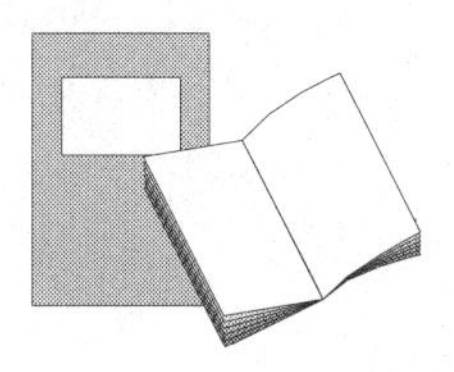

TABLE 9.1 Describing Quantitative and Qualitative Data

Type of data	Central or typical value	Variation or spread in data
Quantitative data: data based on measurements or counts with a standard scale	Mean	Range
Qualitative data: data based on descriptive observations using a nonstandard scale	Median Mode	Frequency distribution

WRITING ABOUT QUANTITATIVE DATA

In Chapter 8 you made a data table to communicate the effect of various concentrations of Chemical X on tomato plant height (see Table 9.3 *The Effect of Various Concentrations of Chemical X on Height of Tomato Plants*). In the data table, you displayed the mean height of the tomato plants. You also calculated the range, a measure of the spread or variation in the height of the tomato plants. To make the data table and to write a title, you had to use your knowledge of the independent variable (concentration of Chemical X) and the dependent variable (height of plants). After you finished the data table, you probably compared the data with your hypothesis. Without realizing it, you began the process of organizing your thoughts to write a paragraph of results. To write a paragraph, follow these steps, one at a time.

Step 1: Write a *topic sentence* stating the independent and dependent variables, and a reference to tables and graphs.

Step 2: Write sentences comparing the *measures of central tendency* of the groups.

Step 3: Write sentences describing the *variation* within the groups.

Step 4: Write sentences stating how the data *support the hypothesis.*

Step 1: Write a Topic Sentence

To begin your paragraph about Chemical X and tomato plants, write a topic sentence about the data table. In the sentence, state the independent and dependent variables and make a reference to appropriate tables and graphs. Some examples of topic sentences are:

- The effect of various amounts of sugar on the activity of mice is depicted in Table X;
- Table Y shows the effect of various types of water pollutants on the walking speed of a water strider.

As you can tell from these examples, there is more than one way to write a topic sentence as long as all the needed information is communicated. Write a topic sentence for the tomato plant data displayed in Table 9.3. When you have finished, compare your topic sentence with the examples in Table 9.4 *Writing about Quantitative Data.*

TABLE 9.2 How to Communicate Descriptive Statistics

Results section of a scientific paper
Quantitative Data: *Write a paragraph that includes a topic sentence, comparison of the means, description of the variation, and support for the hypothesis. Use four steps to guide your writing.* 1. Write a topic sentence stating the independent and dependent variables, and a reference to tables and graphs. 2. Write sentences comparing the measures of central tendency (means) of the groups. 3. Write sentences describing the variation within the groups. 4. Write sentences stating support of the hypothesis by the data.
Qualitative Data: *Determine the type of qualitative data, nominal or ordinal. Then write a paragraph that includes a topic sentence, comparison of the measures of central tendency, description of the variation, and support for the hypothesis. Again, use four steps to guide your writing.* 1. Write a topic sentence stating the independent and dependent variables, and a reference to tables and graphs. 2. Write sentences comparing the measures of central tendency (mode or median) of the groups. 3. Write sentences describing the variation within the groups. 4. Write sentences stating how the data support the hypothesis.
Conclusion section of a scientific paper
Describe the purpose, major findings, an explanation for the findings, and recommendations for further study. Use six questions to guide how you write the conclusion. 1. What was the purpose of the experiment? 2. What were the major findings? 3. Was the hypothesis supported by the data? 4. How did your findings compare with other researchers? 5. What possible explanations can you offer for the findings? 6. What recommendations do you have for further study and for improving the experiment?

TABLE 9.3 The Effect of Various Concentrations of Chemical X on Height of Tomato Plants

Descriptive information	Concentration of Chemical X			
	0%	10%	20%	30%
Mean	15.3 cm	18.1 cm	10.5 cm	6.0 cm
Range	7.0	6.0	6.0	4.0
Maximum	19.0	20.0	14.0	8.0
Minimum	12.0	14.0	8.0	4.0
Number	10	10	10	10

TABLE 9.4 Writing about Quantitative Data

Steps	Examples	Examples
1. Write a **topic sentence** stating the independent and dependent variables, and a reference to tables and graphs.	The effect of various concentrations of Chemical X on the height of tomato plants is summarized in Table 9.3.	Table 9.3 shows the effect of various concentrations of Chemical X on the growth of tomato plants.
2. Write sentences comparing the **measures of central tendency (means)** of the groups.	The mean height of plants grown at 10% X (18.1 cm) was higher than the control (15.3 cm). At higher concentrations of Chemical X, mean plant height was reduced, e.g., 10.5 cm at 20% X and 6.0 cm at 30% X.	The mean height of plants decreased as the concentration of Chemical X increased, e.g., 18.1 cm, 10.5 cm, 6.0 cm. Interestingly, greater mean height was observed at 10% X (18.1 cm) than in the control group (15.3).
3. Write sentences describing the **variation (range or standard deviation)** within the groups.	The range in plant height in the control was greater than the groups receiving 10% X, 20% X and 30% X. As the concentration of Chemical X increased, the range in plant height decreased.	The control group, 10% X group, and 20% X group showed similar variations in height. Less variation occurred at 30% X (Range = 4 cm).
4. Write sentences stating how the data **support the hypothesis.**	The data partially supported the hypothesis that plant growth would decrease as concentration of Chemical X increased. Below 10% X little growth differences were observed; however, at higher concentrations, growth was severely retarded.	At 10% X, growth was not retarded. Above 10% X, plant growth was progressively retarded; thus, the hypothesis was partially supported that growth of plants would decrease as concentrations of Chemical X increased.

Step 2: Compare Measures of Central Tendency

Next, write one or more sentences that compare the means or typical values of the experimental groups. Some examples are:

- The mean mass (150 g) of Elodea grown under blue light was higher than the mean mass (50 g) of the control;
- Water striders have a mean speed of 5 cm/min on unpolluted water as compared with mean speeds of 4 cm/min on water polluted with gasoline and 2 cm/min on water polluted with diesel fuel.

Review the data on mean height of tomato plants in Table 9.3. How do the mean heights compare? Share your comparison with a friend. Now, write sentences describing how the various concentrations of Chemical X affected the height of tomato plants. When you have finished, look at the examples in Table 9.4. How are your sentences different? similar?

Step 3: Describe Variation Within Data

Variation within a set of data is also important to communicate. A group of plants that vary in size from 2 to 12 cm is different from a group of plants that vary in size from 7 to 9 cm, even if the two groups have identical means of 8 cm. The first group has a large range (10 cm) or variation in plant height, while the second group has a small range (2 cm) with little variation in plant height. Some examples of sentences about the variation within data are:

© 1998 PhotoDisc, Inc.

- Rats exposed to high levels of sugar had a small range of 0.5 turns/sec, while those in the control and low sugar groups exhibited a large range of 8 turns/sec;
- Plants in the experimental and control groups exhibited similar variations in size, ranging from 15 to 18 cm in height.

Again, remember that the sentences can be written in many different forms and still be acceptable. Write sentences comparing the height variation among the tomato plants exposed to various concentrations of Chemical X. How do your sentences compare with the examples in Table 9.4 *Writing about Quantitative Data*?

Step 4: State Support for Hypothesis

Conclude the paragraph with a sentence stating how the data support the hypothesis such as:

- Because the mean height of plants receiving fertilizer was greater than the control group, the hypothesis that plant growth would be increased by fertilizer was fully supported;
- Because the mean heights of plants receiving various amounts of compost were equivalent to the control group, the hypothesis that . . . was not fully supported.

After you write the sentence, compare your sentence with the example in Table 9.4.

Finally, construct a paragraph by combining the sentences that resulted from completing the four steps. Is your meaning clear? If not, improve the paragraph by using more effective transition phrases, maintaining the same tense, and so on. As a final step, edit the paragraph for grammar and spelling.

WRITING ABOUT QUALITATIVE DATA

The same process used to write a paragraph about quantitative data can be used to write a paragraph about qualitative data. You just adjust the questions to reflect the appropriate measures of central tendency (modes or medians) and measures of variation (frequency distribution). See Table 9.2 *How to Communicate Descriptive Statistics.*

NOMINAL DATA

In Chapter 8 you made a data table to display the influence of Chemical X on the health of tomato plants. The tomato plants were placed in one of two categories, healthy or unhealthy. This dependent variable is an example of nominal data; that is, data in which items are named and placed in categories. An example of a data table and paragraph of results about the health of tomato plants (nominal data) is provided in Table 9.5 *Writing about Nominal Data.*

ORDINAL DATA

In Chapter 8 you also made a data table to communicate the influence of Chemical X on leaf quality. Leaves were described using a rating from 1 to 4. Because leaf quality is ordinal data, you would compare the medians of the groups. As with the data on plant health, the variation in leaf quality would be described through the frequency distribution. Examples of a data table and paragraph for ordinal data are provided in Table 9.6 *Writing about Ordinal Data.*

TABLE 9.5 Writing about Nominal Data

Descriptive information	Concentration of Chemical X			
	0%	**10%**	**20%**	**30%**
Mode	Healthy	Healthy	Unhealthy	Unhealthy
Frequency distribution				
Healthy	10	8	4	2
Unhealthy	0	2	6	8
Number	10	10	10	10

Steps	Examples
1. Write a **topic sentence** stating the independent and dependent variables, and a reference to tables or graphs.	The influence of Chemical X on the health of tomato plants is summarized in Table 9.5.
2. Write sentences comparing the **measures of central tendency (modes)** of the groups.	When plants received 10% or less Chemical X, they remained healthy. At concentrations of 20% X or above, plant health deteriorated.
3. Write sentences describing the **variation (frequency distribution)** within the groups.	The greatest variation in plant health was at 20% X, where 4 healthy and 6 unhealthy plants occurred. At concentrations below 20% X, plants were predominantly healthy; at 30% X, the majority of plants were unhealthy.
4. Write sentences stating how the data **support the hypothesis**.	The data on plant health supported the hypothesis that higher concentrations of Chemical X would adversely affect plant growth.

TABLE 9.6 Writing about Ordinal Data

Descriptive information	Concentration of Chemical X			
	0%	10%	20%	30%
Median	4	4	2	1
Frequency distribution				
Quality 4	10	6	0	0
Quality 3	0	3	3	0
Quality 2	0	1	7	3
Quality 1	0	0	0	7
Number	10	10	10	10

Steps	Examples
1. Write a **topic sentence** stating the independent and dependent variables, and a reference to tables or graphs.	Leaf quality of plants exposed to various concentrations of Chemical X is summarized in Table 9.6.
2. Write sentences comparing the **measures of central tendency (medians)** of the groups.	High quality leaves, with a rating of 4, were typically found on both control and 10% X plants. At higher concentrations, leaf quality deteriorated, with leaf quality ratings of 2 and 1 on plants grown with 20% X and 30% X.
3. Write sentences describing the **variation (frequency distributions)** within the groups.	The greatest variation in leaf color occurred at 10% X where leaf quality ratings of 4, 3, and 2 were assigned. No variation occurred in the control; all leaves received a rating of 4. At 20% X and 30% X, the plants fell into only two categories, 3 and 2, or 2 and 1.
4. Write sentences stating how the data **support the hypothesis**.	Data on leaf quality supported the hypothesis that Chemical X would impede plant growth.

WRITING A CONCLUSION

In Chapter 6, *Writing Simple Reports,* you used six questions as a guide for writing a conclusion.

- What was the purpose of the experiment?
- What were the major findings?
- Did the data support the hypothesis?
- How did your findings compare with other researchers?
- What possible explanations can you offer for your findings?
- What recommendations do you have for further study and for improving the experiment?

You can also use these steps to write a conclusion about the effect of Chemical X on the growth of tomato plants. As you respond to the questions, you will detect a major difference with Question 2, "What were the major findings?" Previously, you have written about experiments with only one dependent variable, such as time for effervescent tablets to dissolve. Now, you will need to combine information about several dependent variables such as plant height, plant health, and leaf quality. Otherwise, the process is the same.

Write a conclusion for the experiment about the effect of various concentrations of Chemical X on the growth of tomato plants. When you have finished, compare your paragraph with the sample in Table 9.7 *Writing a Conclusion.*

EVALUATING YOUR PARAGRAPHS OF RESULTS AND CONCLUSION

Previously, you learned to use checklists to evaluate a variety of skills, such as analyzing an experimental design diagram and making a graph. Similarly, you can use the checklist in Table 9.8 *Checklist for Evaluating a Paragraph of Results* to evaluate your paragraph of results. Notice that the checklist will work for a paragraph about any type of data, quantitative or qualitative. After you have reviewed your own work, ask a friend or family member to give you feedback. Use their suggestions to improve your work before submitting it to the teacher.

Next, evaluate the conclusion for the experiment using the checklist provided in Table 9.9 *Checklist for Evaluating a Conclusion.* The items on this checklist will be familiar because they are the same that you previously used with a simple report. As always, use the checklist to identify strengths and weaknesses and to make improvements.

Related Web Site

http://www.mste.uiuc.edu/stat/stat.html

TABLE 9.7 Writing a Conclusion

Questions	Examples
1. What was the purpose of the experiment?	The purpose of this experiment was to investigate the effect of various concentrations of Chemical X on the growth of tomato plants.
2. What were the major findings?	At successively higher concentrations of Chemical X, the mean height of the tomato plants decreased and plant health and leaf quality deteriorated. The mean plant height at 10% X (18.1 cm) was greater than the control (15.3 cm), with plants exhibiting similar health. More plants exhibited poor leaf quality in the 10% X group than in the control.
3. Did the data support the hypothesis?	In general, the research data supported the hypothesis that growth of the tomato plants would decrease as the concentration of Chemical X increased.
4. How did your findings compare with other researchers?	Although Crook and Bolton reported that concentrations of 10% X were harmful to radish plants, slightly reduced leaf quality was the only indicator of an adverse effect in this experiment.
5. What possible explanations can you offer for your findings?	Discrepancies in findings could result from different plant species or methods of application. In Crook and Bolton's study, the solution was poured on both plant and soil, whereas in this study the solution was poured only on the soil.
6. What recommendations do you have for further study and for improving the experiment?	Additional studies could be conducted to determine the effect of Chemical X within the 0–20% range on both types of plants and with different methods of application.

TABLE 9.8 Checklist for Evaluating a Paragraph of Results

Criteria	Self	Peer/Family	Teacher
Topic Sentence			
Comparison of measures of central tendency			
Description of variation			
Support of hypothesis by data			
Sentence/Paragraph structure			
Grammar/Spelling			

TABLE 9.9 Checklist for Evaluating a Conclusion

Criteria	Self	Peer/Family	Teacher
Purpose of experiment			
Major findings			
Support of hypothesis by data			
Comparison with other research			
Explanations for findings			
Recommendations			
Sentence structure			
Grammar/Spelling			

USING TECHNOLOGY

Integrated Software Programs

In Chapter 8, you conducted an experiment to determine how minerals in water affect suds formation. Use your knowledge of integrated software programs, such as ClarisWorks, Microsoft Office, or WordPerfect, to write a report for "A Sudsy Experience." Expand the simple report, described in Chapter 6, to include summary data tables, graphs, and paragraphs of results.

Specific components of the integrated software package you might use are described below.

Word Processing: general text of document
Drawing a Table: experimental design diagram
Spreadsheet: data tables and graphs/charts
Internet Access: background information for the introduction and conclusion.

CHAPTER 10

Displaying Dispersion/ Variation in Data

Objectives

- Make stem-and-leaf plot(s) for a set of data and write a paragraph to summarize findings.
- Make a boxplot for a set of data, construct a data table, and write a paragraph to summarize findings.
- Calculate the variance/standard deviation for a sample and make inferences to a normally distributed population; make a graphical display and data table and write a paragraph to summarize findings.
- Evaluate a set of data and determine the most appropriate measure(s) of dispersion/variation to use.

National Standards Connections

- Use appropriate tools and techniques to gather, analyze, and interpret data (NSES).
- Use technology and mathematics to improve investigations and communications (NSES).
- Understand and apply measures of central tendency and variability (NCTM).

In Chapter 8, *Analyzing Experimental Data*, you learned how to summarize experimental data using measures of central tendency and variation. You used simple measures of variation such as range and frequency distribution. In this chapter several techniques are presented to help you expand the possibilities for displaying the dispersion or variation within sets of quantitative data. New techniques include stem-and-leaf plots, boxplots (box and whisker diagrams), and standard deviation/variance. First, you will learn these methods using pencil and paper calculations; then, if graphing calculators are available, you will use them to calculate statistics and to display the data. With these additional techniques, you will better understand dispersion/variance in data and will be more able to make informed decisions about differences among groups of experimental data. These new skills will also establish a foundation for applying the inferential statistics described in Chapter 11.

OVERVIEW OF STATISTICAL AND GRAPHICAL PRESENTATIONS

Let's begin by comparing displays of tomato production by 53 plants in a garden. By comparing the three types of data displays you can determine the pros and cons of each type of display. Below is the number of tomatoes produced by each of the 53 garden plants.

Number of tomatoes produced
8, 17, 23, 35, 45, 56, 65, 94, 32, 31, 14, 15, 54, 91, 18, 15, 28, 9, 29, 30, 39, 35, 47, 48, 69, 8, 14, 19, 22, 42, 22, 29, 7, 28, 24, 30, 35, 44, 62, 17, 23, 15, 35, 34, 29, 32, 39, 45, 53, 35, 42, 57, 94.

STEM-AND-LEAF PLOTS

Stem-and-leaf plots are a form of bar graph where numeric data are plotted by using the actual numerals in the data to form the graph. They are a quick way to display 25 or more pieces of data. Each piece of data is displayed as two parts: a **stem** and a **leaf.**

In a stem-and-leaf plot, a number, such as 24, is displayed as 2|4; 2 is the stem and 4 is the leaf. Several numbers, such as the tomato data in the 50's (56, 54, 53, 57) would be displayed as 5|6 4 3 7. When ordered, the stem and leaves would appear as 5|3 4 6 7. Include a key, such as 4|8 = 48, as an aid to interpreting the values that the stems and leaves represent.

Data such as the number of tomatoes produced are first put in an **unordered** stem-and-leaf plot (Figure 10.1), which is later converted to an **ordered** stem-and-leaf plot (Figure10.2). If the data are to be grouped by tens, begin by listing the tens digits in order and drawing a line to the right. These form the **stem** of the graph. Next, go through the tomato data and write the ones digit next to the appropriate tens digit. These are the **leaves.** Stem-and-leaf plots can also be used for larger data. For data in the 100's, for example, the stem is the hundreds digit and the leaves are the tens and ones, such as 463, 479, 482 represented as 4|63, 79, 82. Commas or extra space can be used to separate each leaf.

Key: 4|8 = 48

Stem	Leaf
0	8 9 8 7
1	7 4 5 8 5 4 9 7 5
2	3 8 9 2 2 9 8 4 3 9
3	5 2 1 0 9 5 0 5 5 4 2 9 5
4	5 7 8 2 4 5 2
5	6 4 3 7
6	5 9 2
7	
8	
9	4 1 4

Figure 10.1 Unordered Stem-and-Leaf Plot.

Key: 4|8 = 48

Stem	Leaf
0	7 8 8 9
1	4 4 5 5 5 7 7 8 9
2	2 2 3 3 4 8 8 9 9 9
3	0 0 1 2 2 4 5 5 5 5 5 9 9
4	2 2 4 5 5 7 8
5	3 4 6 7
6	2 5 9
7	
8	
9	1 1 4

Figure 10.2 Ordered Stem-and-Leaf Plot.

From the stem-and-leaf plots, answer the following questions about tomato production by various garden plants. Then, check your answers.

1. What values are represented by 4|8, 9|1, 0|8?
2. What was the minimum number of tomatoes produced by a plant? the maximum number? the range in tomato production?
3. What was the most frequent number (mode) of tomatoes produced?
4. Are there **gaps** or empty places in the data?
5. Are there **clusters** or isolated groups of data?
6. What is the general shape of the distribution?

If you are like most students, you will find that stem-and-leaf plots are much easier to construct than bar graphs. In addition, they provide an efficient method of ordering data where individual pieces of data are easily identified. Note in Figures 10.3 a, b, c, and d how effectively stem-and-leaf plots illustrate various data distribution patterns: Bell-shaped (Normal), U-shaped, J-shaped, and Rectangular-shaped.

Answers to Questions on page 106. 1) 48, 91, 8. **2)** Minimum = 7; maximum = 94; range = 87. **3)** Mode = 35. **4)** Gaps in 70's and 80's. **5)** Clusters = 91, 94, 94 **6)** Bell-shaped except for cluster at high end.

a. Bell-shaped (Normal) Distribution: highs balance the lows.

1 | 0
2 | 3 3 8
3 | 2 5 6 7
4 | 3 3 4 5 5 8 8 9
5 | 2 5 5 8 9
6 | 4 2 7
7 | 2 3
8 | 1
9 |

b. U-shaped (Bi-Modal) Distribution: two groups of relatively equal frequency; if studied separately, each group may be bell-shaped.

1 | 0 2 2 7 8 9
2 | 3 4 5 6 6
3 | 0 4 5
4 | 2 4
5 | 1 6 7
6 | 2 3 4 5 8
7 | 1 1 2 5 7 8 8
8 |
9 |

c. J-shaped Distribution: shows that there is probably a limit to the values of the data.

1 | 0 2 2 4 5 7 7 8 9
2 | 1 3 3 6 7 7
3 | 0 4 5 6 8
4 | 3 4 5 3
5 | 2 3 4
6 | 7 8
7 | 2
8 | 2
9 | 0

d. Rectangular-shaped distribution: uniform distribution of data with values evenly distributed over a range.

1 |
2 |
3 | 0 3 3 8 4
4 | 2 3 4 4 7 8
5 | 1 2 4 8 8
6 | 3 4 5 5 9 9
7 | 2 3 4 7 7
8 |
9 |

Figure 10.3 Common Data Distribution Patterns.

BOXPLOTS (BOX AND WHISKER DIAGRAMS)

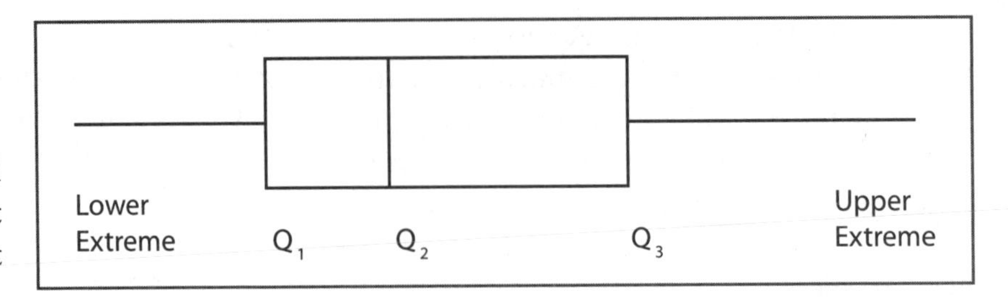

A boxplot is a graphical method of displaying data that is based upon five important components of the data.

Lower extreme	Minimum value
Lower quartile (Q_1)	Number below which 25% of the values fall
Median (Q_2)	Number which divides the data into half, with 50% of the values falling above the number and 50% below
Upper quartile (Q_3)	Number below which 75% of the numbers fall
Upper Extreme	Maximum value

As shown above, each of these components is displayed by a specific part of the boxplot.

Use this boxplot of tomato production to answer the questions that follow (see Figure 10.4).

1. What is the lower extreme? Upper extreme?
2. What is $Q_{1,}$ Q_2 (median), Q_3?
3. Below which value do 25% of the plants produce? 50% of the plants? 75% of the plants?
4. In what range is the middle 50% of production? Are these middle values for tomato production symmetrically distributed? How can you tell? (Hint: Look at the box.)
5. Where do you have the greatest dispersion or spread in data: the lower quartile or the upper quartile? How can you tell? (Hint: Look at the length of the lines or whiskers.)
6. Because you were estimating from a graph, your numbers may not be exact. How could you use the stem-and-leaf plot to obtain more specific information?

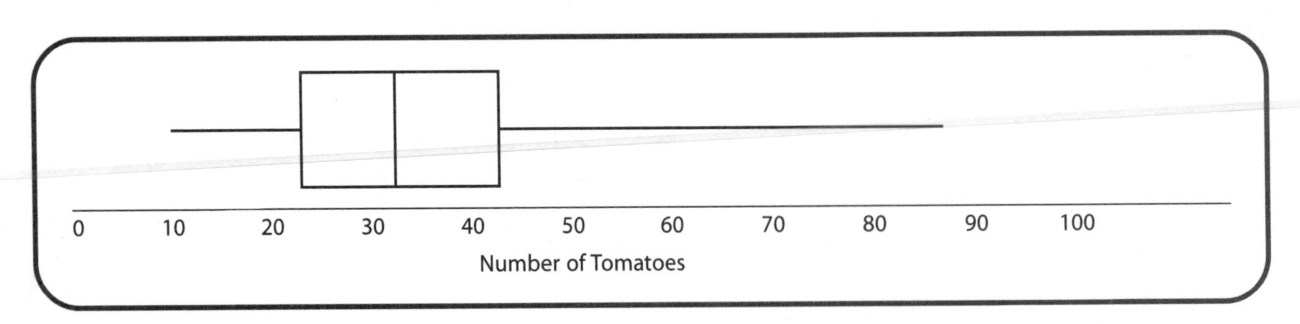

Figure 10.4 Boxplot of Number of Tomatoes Produced.

Answers to Question about Figure 10.4: 1) Lower extreme = 7; upper extreme = 94. **2)** Q_1 = 20.5; Q_2 (median) = 32; Q_3 = 45. **3)** 35% are below 20.5; 50% are below 32, and 75% are below 45. **4)** 50% are between 20.5 and 45; values are fairly symmetrical because the two halves of the box are approximately the same lengths. **5)** Greatest dispersion is in the upper quartile where the line (whisker) is much longer. **6)** From the stem-and-leaf plot, you can determine the median. Then, you can determine the median for the lower half (Q_1) and the median for the upper half (Q_3). See if you can do this. How do your answers compare with ours?

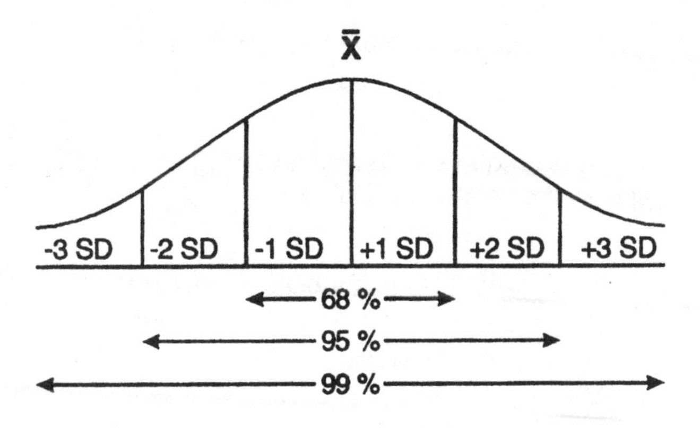

Figure 10.5 Normal Distribution of a Population.

STANDARD DEVIATION AND VARIANCE

The standard deviation is a powerful measure of dispersion that is frequently reported in scientific papers. Dispersion is calculated by comparing each individual piece of data in the sample with the mean of the sample. These differences are entered into a specific mathematical formula that is solved to yield the standard deviation. The greater the standard deviation, the more variation or dispersion in the data.

Typically, scientists use the standard deviation when they want to make inferences from a sample to a population. The **sample** is the actual set of objects measured or described; for example, the 53 plants on which the tomatoes were counted. The **population** is the set of "all" tomato plants of the same type grown in similar gardens.

Scientists and statisticians generally assume that populations follow a normal distribution that is illustrated by a bell-shaped curve such as the one shown in Figure 10.5. In a normally distributed population, the mean is found at the peak of the curve. Sixty-eight percent (68%) of the population falls within ± 1 standard deviation of the mean. Ninety-five percent (95%) of the population falls within ± 2SD of the mean. Similarly, 99% of the population falls within ± 3 standard deviations of the mean. Populations frequently consist of very large numbers. For a sample to reflect a population, it must also be sufficiently large. Typically, statisticians recommend that sample sizes be a minimum of 30.

When the standard deviation of a population is graphically presented in this text, the following symbols will be used. In other text, or on graphing calculators, symbols may vary; because a standard system, such as used with boxplots, does not exist. (See Figure 10.6, *Graphic Display of Standard Deviations of the Entire Population.*)

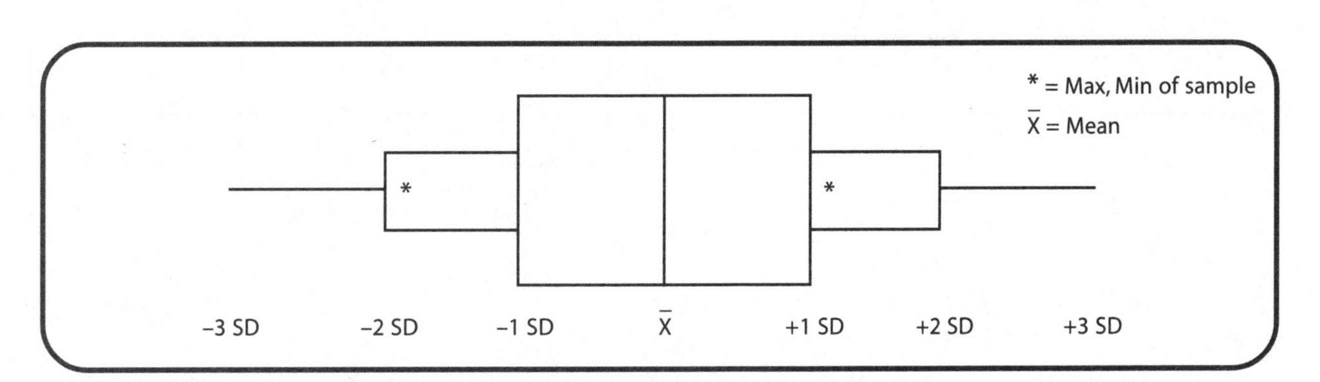

Figure 10.6 Graphic Display of Standard Deviations of the Entire Population.

Below are a data table and graphical presentation of tomato production by the 53 garden plants (see Table 10.1 and Figure 10.7). Use the data table and graph to answer the questions that follow.

1. What is the mean? Standard deviation?
2. What range of tomato production would you expect for 68% of the plants?
3. What range of tomato production would you expect for 95% of the plants? 99% of the plants? Are the values reasonable? Why?
4. Why does the data table and graphical display include values that were not in the sample?

The tomato production predicted for the population of "all" tomato plants is not reasonable because there are negative values of –6.1 and –26.9 tomatoes. Remember that to predict from a sample to a population, a normal distribution is assumed. Refer back to the stem-and-leaf plot and the boxplot. Although the lower part of the sample appeared "mound shaped," the sample as a whole was not. A gap occurred in the data in the 70's and 80's, with only a few data values occurring in the 90's. The data were **skewed**, or had a tail, at the upper end. You can see this skew in the boxplot because the upper end is substantially longer. When a sample is skewed, stem-and-leaf plots and boxplots are more appropriate ways to display the data. By displaying data in various ways, you can make more informed decisions about the most appropriate way to communicate their findings.

Answers about Standard Deviation/Variance of Tomatoes. 1) Mean = 35.5, SD = 20.8 **2)** 14.7 to 56.3 tomatoes **3)** –6.1 to 77.1 tomatoes; –26.9 to 97.9 tomatoes; values are not reasonable because you cannot have a negative (–) number of tomatoes. **4)** The data table and graph are based on inferences made from the sample to the population and include values that you would expect in the population based upon the dispersion/variation in the sample.

TABLE 10.1 Tomatoes Produced by a Population of Plants

Descriptive information	Number of tomatoes
Mean	35.5
Variance	432.6
Standard deviation	20.8
± 1 SD (68% range)	14.7 to 56.3
± 2 SD (95% range)	–6.1 to 77.1
± 3 SD (99% range)	-26.9 to 97.9
Number	53

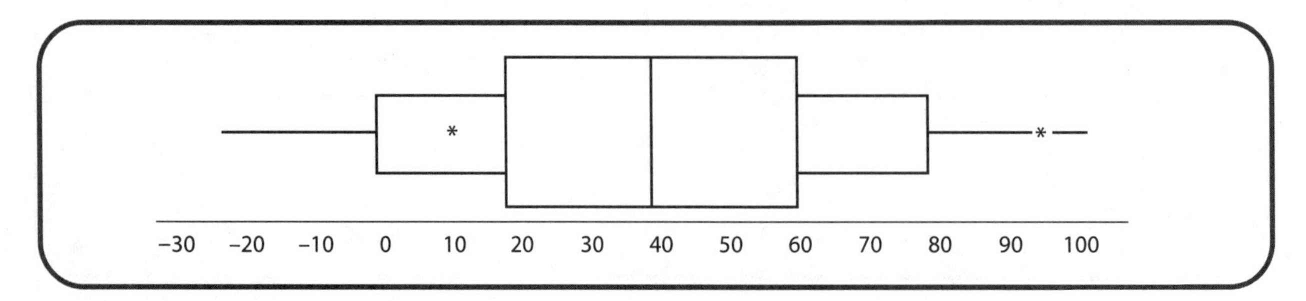

Figure 10.7 Projected Number of Tomatoes Produced by a Population of Plants.

CALCULATING DISPERSION/ VARIATION IN EXPERIMENTAL DATA

Have you ever played with the paper covering on a straw, maybe pushing it tightly together at one end? If so, you know that you get a "worm-like" piece of paper. Use the procedure in Investigation 10.1, *Paper Worms*, to turn these "worms" into an experiment (see page 122).

Below are data on paper worms collected by a class of 30 students (see Table 10.2). The data will be used to calculate various measures of dispersion/variation. Follow these procedures to analyze the data from the paper worm experiment.

TABLE 10.2 Length of Paper Worms (mm)

0 drops of water (dry)						*4 drops of water (wet)*					
30	35	74	49	55	20	65	55	114	89	80	55
38	40	50	65	32	40	85	40	91	105	75	67
24	38	44	40	58	66	60	65	81	90	65	97
40	64	37	34	38	44	57	95	87	90	55	76
48	70	43	49	37	45	85	110	75	70	37	74

INVESTIGATION 10.1 ■ Paper Worms

Question

How do paper worms respond to water?

Materials for Each Student

- Straw with paper wrapper
- Small container of water
- Metric ruler
- Eyedropper or straw

Safety

- Dispose of straws after experimenting.
- Do not place straws in mouth.

Procedure

1. Hold a straw vertically. Tear the paper at the top end of the straw. Carefully, push the paper tightly together down to the bottom end. Try not to tear the paper. When you have finished, remove the paper worm.
2. Measure the length of the paper worm in millimeters (mm). If the paper worm is curved, measure the shortest distance between the two ends. Record the data.
3. Using an eyedropper (or your straw), add 4 drops of water to the worm. Measure the length of the worm (mm). Record your data.

	0 drops water (dry)	**4 drops water (wet)**
Length (mm)	______	______

4. Enter the data in a class data table. If necessary, repeat the experiment so that your class has 30 or more trials.

Analyzing the Data

1. Make stem-and-leaf plots of both the wet and dry paper worm data. Write a paragraph describing the data.
2. Make boxplots of both the data sets. Display the information in a data table and a graph. Write a paragraph about the results.
3. Calculate the standard deviation/variance for a population of wet and dry paper worms. Display the information in both a data table and a graph. Write a paragraph about the results.
4. What are appropriate ways to display your experimental findings? Why?
5. Write a report about your experiment with paper worms.

(continued on the following page)

INVESTIGATION 10.1 ▪ Paper Worms *(continued)*

Extending Your Learning

1. What type of energy did you use to make the paper worm?
   What type of energy did the paper worm have?
   What type of energy did the water possess?
   What energy transformations occurred as you added water to the paper worm?
2. What is the chemical composition of paper? How did the water interact with the chemicals in the paper?
3. How is tissue paper, the type of paper around the straw, manufactured? Do you think you would get the same results with other types of paper? Why?
4. What other factors affect the action of paper worms? Design and conduct experiments to test your hypotheses.

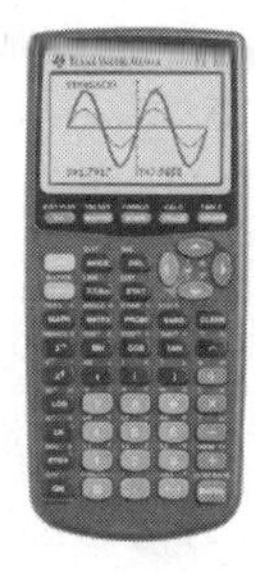

USING TECHNOLOGY

1. In the **STAT** mode of your calculator, enter the dry worm and wet worm measurements in Lists 1 and 2. Sort the measurements in ascending order. Use the ordered data to make ordered stem-and-leaf plots. (See Appendix A, *Using Technology*, for additional help in using the graphing calculator.)
2. In the **STAT** mode select CALC (for calculate) and then 1 VAR (for 1-variable statistics). Depending on the brand of your calculator, you will need to enter the desired list number (e.g., L1), or SET the 1-variable x-list to the desired list number before selecting 1 VAR. *Repeat* the selection process for the wet worm data by changing the list number to List 2. Obtain 1-variable statistics for each list of data including mean, median, quartiles, minimum, maximum, standard deviation, and number. Use this information to make data tables and graphical displays.
3. Set up a different graph for the dry worm and wet worm data sets. In setting up the first graph, select boxplot as your graph type and List 1 for the x-list and 1 for the frequency. Repeat to set up the second graph, but enter List 2 for the x-list. Graph both boxplots simultaneously.
4. Press **Trace** and use the left and right arrow keys to highlight such values as the minimum, 1st quartile, median, 3rd quartile, and maximum. To make comparisons between the two boxplots, use the up or down arrows to toggle between graphs.
5. Link your graphing calculator to a computer and download these boxplots to save, enlarge, or print them.

STEM-AND-LEAF PLOTS

In the beginning of this chapter, a stem-and-leaf plot was made for one set of data. In this experiment, there are two sets of data, one for 0 drops of water and one for 4 drops of water. When two sets of data exist, a back-to-back stem-and-leaf plot can be constructed. If more than two sets of data exist, you can construct multiple stem-and-leaf plots. To facilitate data interpretation, use the same scale for the "stems" and place the plots beside or underneath each other.

Step 1: Determine the minimum and maximum values of the data. For 0 drops of water, the values are 20 mm and 74 mm. For 4 drops of water, the values are 37 mm and 114 mm. In constructing the back-to-back stem-and-leaf plot, use the minimum and maximum values from both data sets, e.g. 20 mm and 114 mm.

Step 2: In the center of the page, write the "stems" vertically. Draw a dark line on both sides of the stem.

Step 3: On the left side of the stem, write the "ones place" values of the data for 0 water drops. On the right side of the stem, write the "ones place" values of the data for 4 water drops. This is an unordered stem-and-leaf plot as shown in Figure 10.8, *Back-to-Back Unordered Stem-and-Leaf Plots for Impact of Water on the Length of Paper Worms.*

Step 4: Make an ordered stem-and-leaf plot by arranging the leaves in order from the "inside" to the "outside." Be sure to include a key (see Figure 10.9, *Back-to-Back Ordered Stem-and-Leaf Plots and Paragraph for Impact of Water on Length of Paper Worms (mm)*).

Step 5: Write a paragraph about the data. In the paragraph provide information, such as the following:

- Minimum and maximum values and range
- Most frequently occurring value (mode)
- Shape-of-the-data
- Existence of clusters and gaps
- Support for hypothesis.

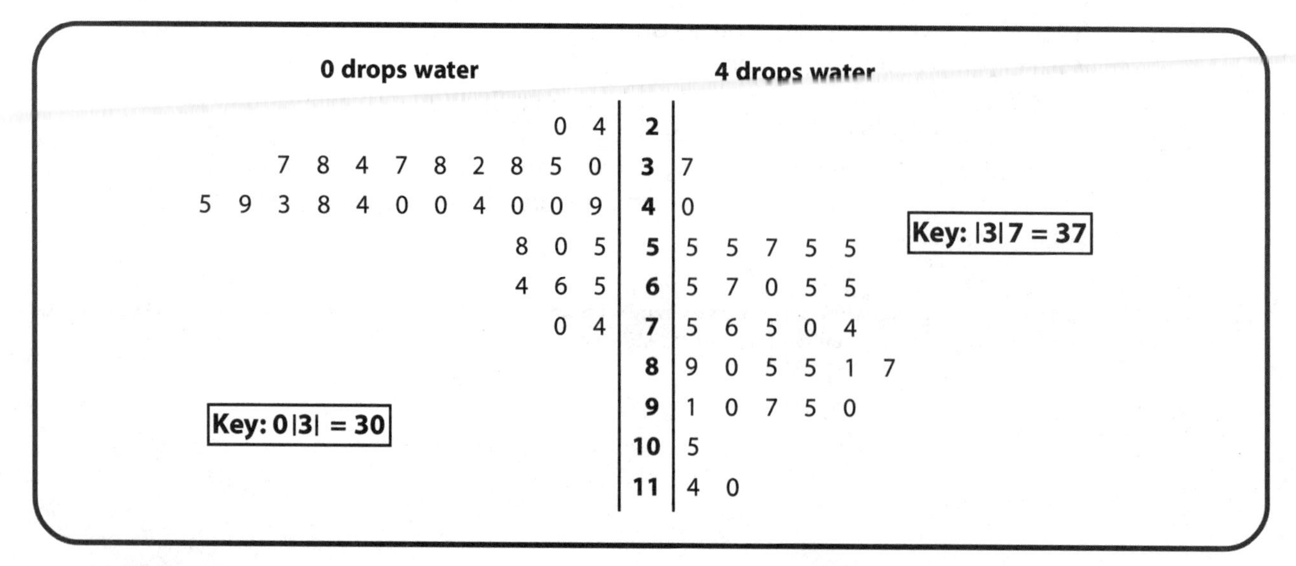

Figure 10.8 Back-to-Back Unordered Stem-and-Leaf Plots for Impact of Water on the Length of Paper Worms.

Stem-and-Leaf Plot

0 drops water		4 drops water
4 0	**2**	
8 8 8 7 7 5 4 2 0	**3**	7
9 9 8 5 4 4 3 0 0 0 0	**4**	0
8 5 0	**5**	5 5 5 7
6 5 4	**6**	0 5 5 5 7
4 0	**7**	0 4 5 5 6
	8	0 1 5 5 7 9
	9	0 0 1 5 7
	10	5
	11	0 4

Key: 0|3| = 30 **Key: |3|7 = 37**

Paragraph

Initially, the length of the paper worms ranged from 20 mm to 74 mm, with the most typical length being 40 mm. After 4 drops of water were added, lengths ranged from 37 mm to 114 mm, with the most typical measurements being 55 mm and 65 mm. For the dry worms, the data formed a very steep mound; most of the measurements were in the 30's and 40's. For wet worms, the data was a much flatter mound, with the center part being rectangular. The data supported the hypothesis that water would increase worm length. Typical lengths shifted from the 20's and 30's to the 50's, 60's, 70's, 80's, and 90's.

Figure 10.9 Back-to-Back Ordered Stem-and-Leaf Plots and Paragraph for Impact of Water on Length of Paper Worms (mm).

BOXPLOTS (BOX AND WHISKER DIAGRAMS)

Previously, you looked at a boxplot of tomato production by various garden plants. Use this experience to help you construct boxplots for the paper worm data. Begin by looking at the ordered stem-and-leaf plots for the dry and wet paper worm data. Then, use the following procedures to make boxplots of your paper worm data. Our data are shown in Figure 10.10, *Ordered Stem-and-Leaf Plots for Dry and Wet Paper Worms.*

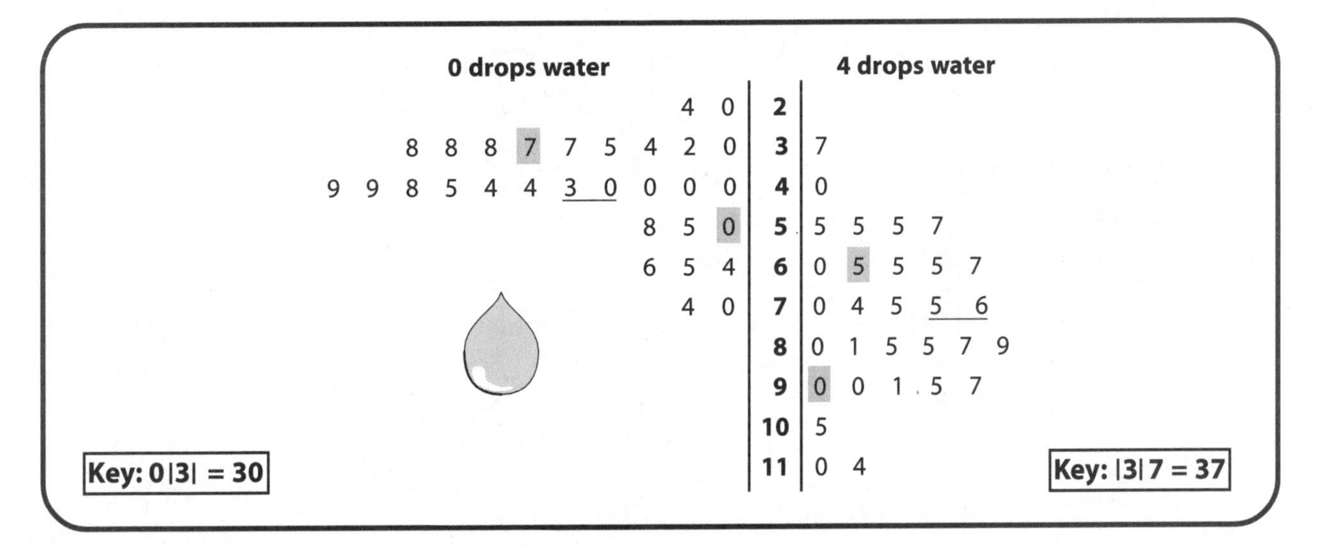

0 drops water		4 drops water
4 0	**2**	
8 8 8 7 7 5 4 2 0	**3**	7
9 9 8 5 4 4 3 0 0 0 0	**4**	0
8 5 0	**5**	5 5 5 7
6 5 4	**6**	0 5 5 5 7
4 0	**7**	0 4 5 5 6
	8	0 1 5 5 7 9
	9	0 0 1 5 7
	10	5
	11	0 4

Key: 0|3| = 30 **Key: |3|7 = 37**

Figure 10.10 Ordered Stem-and-Leaf Plots for Dry and Wet Paper Worms.

Step 1: Find the median or mid-point of the data. Both sets of data contain 30 values. The median will be the point halfway between the 15th and 16th value (see underlined values).

Median (Q_2) for 0 drops $\frac{40 \text{ mm} + 43 \text{ mm}}{2} = 41.5$ mm

Median (Q_2) for 4 drops $\frac{75 \text{ mm} + 76 \text{ mm}}{2} = 75.5$ mm

Step 2: Find the median of the lower half of the data (Q_1). Because 15 values are found in the lower half, Q_1 is the 8th number (see shaded values).

Lower quartile (Q_1) for 0 drops 37 mm

Lower quartile (Q_1) for 4 drops 65 mm

Step 3: Find the median of the upper half of the data (Q_3). Fifteen values are found in the upper half. The median of these values is the 8th number because 7 numbers fall above and below it (see shaded valued).

Upper quartile (Q_3) for 0 drops 50 mm

Upper quartile (Q_3) for 4 drops 90 mm

Step 4: Find the extreme values of the data, that is the minimum and the maximum values.

0 drops Min = 20 mm Max = 74 mm

4 drops Min = 37 mm Max = 114 mm

Step 5: Develop a scale for plotting the values. Use the techniques previously described in Chapter 5.

$\frac{\text{Max} - \text{Min}}{5} = \frac{114 \text{ mm} - 20 \text{ mm}}{5} = \frac{94 \text{ mm}}{5} = 18.8 \text{ mm} \sim 20 \text{ mm}$

Generally, mathematicians graph boxplots on a horizontal number line. Graphing calculators also plot horizontally. Often scientists will plot vertically, so that the independent variable (x) and the dependent variable (y) are on the expected axes. Either method is appropriate as long as the variables are clearly labeled.

Step 6: Plot points for the upper quartile (Q_3) and the lower quartile (Q_1). Make a box. Draw a vertical line across the box at the median (Q_2). Draw a line or "whisker" from each quartile to the extreme values. (See Table 10.2, *Length of Paper Worms (mm)*.)

Step 7: Find the **interquartile ranges (IQR).** This is the difference between Q_3 and Q_1.

0 drops 50 mm – 37 mm = 13 mm

4 drops 90 mm – 65 mm = 25 mm

Step 8: Determine if **outliers**—very small or very large points—exist. Such points are greater or less than 1.5 times the interquartile range.

0 drops IQR = 13 mm 13 mm × 1.5 = 19.5 mm

Lowest reasonable value = Q_1 – 19.5 mm =
37 mm – 19.5 mm = 17.5 mm

Highest reasonable value = Q_3 + 19.5 mm =
50 mm + 19.5mm = 69.5 mm

Outliers exist at the upper end: 70 and 74 mm. No outliers exist at the lower end. Place an asterisk (*) on the whisker where these points exist.

4 drops IQR = 25 mm 25 mm × 1.5 = 37.5 mm

Lowest reasonable value = Q_1 – 37.5 mm

65 mm – 37.5 mm = 27.5 mm

Highest reasonable value = Q_3 + 37.5 mm

90 mm + 37.5 mm = 127.5 mm

No outliers exist.

Step 9: Write a paragraph about the boxplots that includes the following information (see Table 10.3, *Summary Data Table and Boxplots for the Impact of Water on the Length of Paper Worms (mm)*):

- Introductory sentence
- Comparison of medians
- Comparison of boxes and interquartile range (middle 50% of values)
- Symmetry of distribution
- Range of values and outliers
- Support for hypothesis.

TABLE 10.3 Summary Data Table and Boxplots for the Impact of Water on the Length of Paper Worms (mm)

	Number of water drops	
Descriptive information	**0**	**4**
Median (Q_2)	41.5	75.5
Dispersion		
Minimum	20	37
Q_1	37	65
Q_3	50	90
Maximum	74	114
Interquartile range	13	25
Lowest reasonable value	17.5	27.5
Highest reasonable value	69.5	127.5
Outliers	70, 74	None
Number	30	30

Boxplots

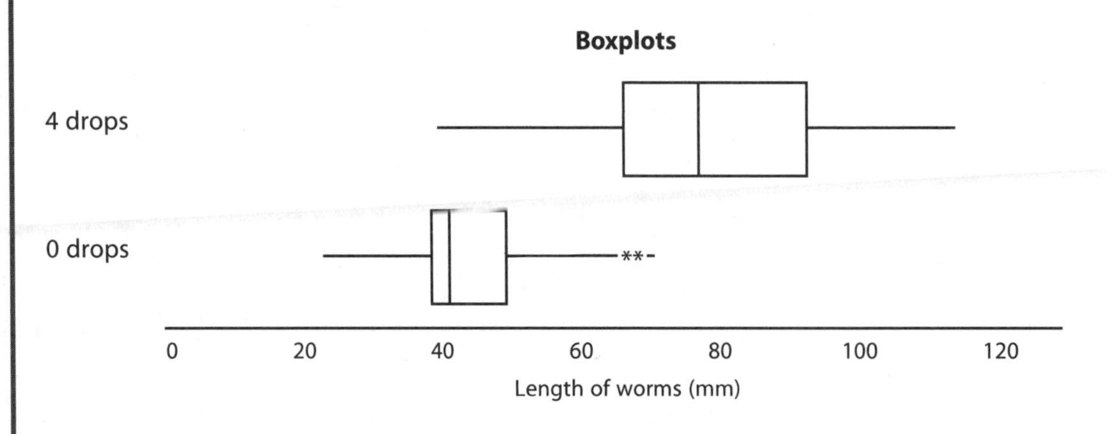

Paragraph

The effect of water on the length of paper worms is displayed in the table above and accompanying boxplots. When water was added, the median length of the paper worms increased from 41.5 mm to 75.5 mm. For both dry and wet worms, the middle 50% of lengths were fairly symmetrical. Almost twice as much dispersion occurred in wet worms as evidenced by the interquartile range: 25 mm versus 13 mm. Overall, wet worms also showed greater dispersion, a total range of 77 mm versus 54 mm. For both sets of worms, dispersion was greatest in the upper quartile. Two outliers, 70 mm and 74 mm, occurred in the dry worm data. The data supported the hypothesis that water would increase worm length, because the medians were approximately two quartiles apart, with the upper 50% of dry worm lengths overlapping the lower 50% of wet worm lengths.

STANDARD DEVIATION AND VARIANCE

Use the following procedures to calculate the standard deviation and variance for the dry paper worm sample, and then the wet paper worm sample. Four basic steps are involved in calculating the variance and standard deviation of a sample. These steps are described below and in Table 10.4 *Calculating the Variance and Standard Deviation of the Dry Worm Lengths (mm)* for the dry paper worm data.

Step 1: Find the difference between each individual value (X_i) and the mean ($\bar{X}$).

Examples
- Value 1: 20 mm – 44.9 mm = –24.9 mm
- Value 19: 45 mm – 44.9 mm = 0.1 mm
- Value 29: 70 mm – 44.9 mm = 25.1 mm

Step 2: Square the difference between each individual value and the mean.

Examples
- Value 1: $(-24.9 \text{ mm})^2 = 620.01 \text{ mm}^2$
- Value 19: $(0.1 \text{ mm})^2 = .01 \text{ mm}^2$
- Value 29: $(25.1 \text{ mm})^2 = 630.01 \text{ mm}^2$

Step 3: Find the sum of the squared differences $(X_i - \bar{X})^2$. Divide this difference by the degrees of freedom, that is the number in the sample (n) minus 1. The dividend is the unbiased estimate of the variance of the population.

$$\text{Variance} = \frac{620.01 \text{ mm}^2 + 436.81 \text{ mm}^2 \ldots\ldots + 846.81 \text{ mm}^2}{30 - 1}$$

$$= \frac{5064.7 \text{ mm}^2}{29}$$

$$= 174.6 \text{ mm}^2$$

Step 4: Find the square root of the variance. This number is the unbiased estimate of the standard deviation of the population.

$$\text{Standard deviation} = \sqrt{174.6 \text{ mm}^2} = 13.2 \text{ mm}$$

Statisticians use a biased and unbiased measure of the standard deviation of the population. A **biased** measure is found by dividing by the number (n) in the sample; it is generally too low an estimate of the population estimate. The **unbiased estimate** is found by dividing by the degrees of freedom (n – 1). In this text, the unbiased estimate of variance and standard deviation will be used. On calculators, this generally has the following symbols: Texas Instruments (Sx or σ n – 1) and Casio (x σ n – 1).

TABLE 10.4 Calculating the Variance and Standard Deviation of the Dry Worm Lengths (mm)

Trial	Individual Values (X_i)	Mean ($\bar{X}$)	Step 1 Individual Value – Mean ($X_i - \bar{X}$)	Step 2 Squared difference $(X_i - \bar{X})^2$
1	20	44.9	–24.9	620.01
2	24	44.9	–20.9	436.81
3	30	44.9	–14.9	222.01
4	32	44.9	–12.9	166.41
5	34	44.9	–10.9	118.81
6	35	44.9	–9.9	98.01
7	37	44.9	–7.9	62.41
8	37	44.9	–7.9	62.41
9	38	44.9	–6.9	47.61
10	38	44.9	–6.9	47.61
11	38	44.9	–6.9	47.61
12	40	44.9	–4.9	24.01
13	40	44.9	–4.9	24.01
14	40	44.9	–4.9	24.01
15	40	44.9	–4.9	24.01
16	43	44.9	–1.9	3.61
17	44	44.9	–0.9	0.81
18	44	44.9	–0.9	0.81
19	45	44.9	0.1	0.01
20	48	44.9	3.1	9.61
21	49	44.9	4.1	16.81
22	49	44.9	4.1	16.81
23	50	44.9	5.1	26.01
24	55	44.9	10.1	102.01
25	58	44.9	13.1	171.61
26	64	44.9	15.1	364.81
27	65	44.9	20.1	404.01
28	66	44.9	21.1	445.21
29	70	44.9	25.1	630.01
30	74	44.9	29.1	846.81

Mean = $\frac{1347 \text{ mm}}{30}$ = 44.9 mm

Step 3 Variance = $\frac{5464.7 \text{ mm}^2}{30 - 1}$ = 174.6 mm^2

Step 4 SD = $\sqrt{174.6 \text{ mm}^2}$ = 13.2 mm

TABLE 10.5 The Effect of Water on the Standard Deviation/Variance of Paper Worm Lengths (mm)

Data Table

Descriptive information	Number of water drops	
	0	**4**
Mean	41.5	76.3
Standard deviation (unbiased)	13.2	19.1
± 1 SD (68% range)	31.7 to 58.1	57.2 to 94.5
± 2 SD (95% range)	18.5 to 71.3	38.1 to 114.5
± 3 SD (99% range)	5.3 to 84.5	19.0 to 133.6
Number	30	30

Graphic Display

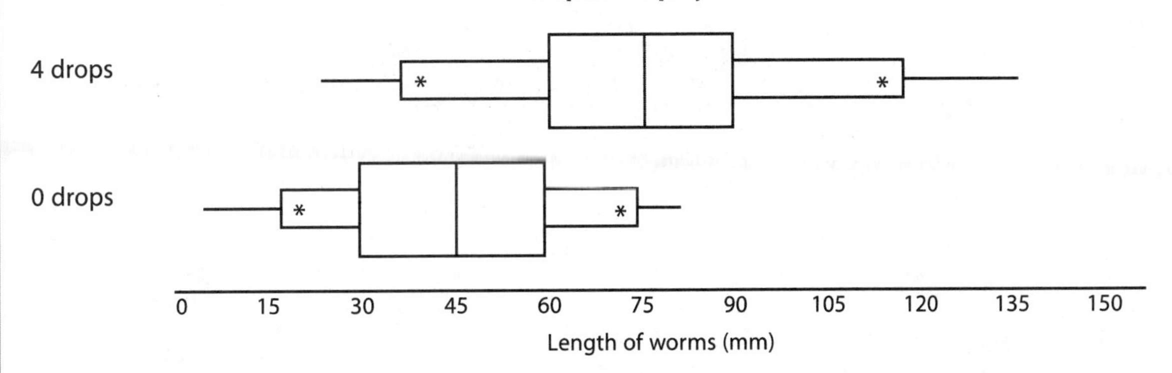

Paragraph

The impact of water on the length of paper worms is displayed in the above data table and graph. After water was added, the mean length of the worms increased from 44.9 mm to 76.3 mm. Greater dispersion occurred in the wet paper worms. Projections for the populations of "all" worms showed that the means were approximately 2 SD apart. The data supported the hypothesis that water would increase the length of the worms.

MAKING DECISIONS ABOUT DISPERSION/VARIATION IN DATA

In Chapter 8, you used a table to make decisions about appropriate measures of central tendency and dispersion. You began by classifying the independent variable as continuous or discrete. Then, you classified the dependent variable as quantitative (continuous/discrete), qualitative (ordinal), or qualitative (nominal). Table 10.6 *Making Decisions about Measures of Central Tendency and Dispersion/Variation* is an expanded version of the table that includes the new measures of dispersion/variation and graphical displays that you learned about in this chapter.

Use the information in Table 10.6 to determine appropriate statistics for Investigation 10.2, *Watery Statistics*. Then, complete the investigation and write a report. You may also find the table helpful in solving the practice problems at the end of this chapter.

TABLE 10.6 Making Decisions about Measures of Central Tendency and Dispersion/Variation

Levels of Independent Variable	Dependent Variable	Central Tendency	Dispersion/ Variation	Graphic Displays
Continuous	Quantitative (continuous/ discrete)	Mode Median Mean	Range Variance Standard Deviation	Bar Graph Line Graph, Line-of-Best Fit (continuous) Line Graph, Broken Line (discrete) Stem-and-Leaf Plots Boxplots Standard Deviation
	Qualitative (ordinal)	Mode Median	Frequency distribution	Bar Graph (medians) Frequency distribution
	Qualitative (nominal)	Mode	Frequency distribution	Frequency distribution
Discrete	Quantitative (continuous/ discrete)	Mode Median Mean	Range Variance Standard Deviation	Bar Graph Stem-and-Leaf Plots Boxplots Standard Deviation
	Qualitative (ordinal)	Mode Median	Frequency distribution	Bar graph (medians) Frequency distribution
	Qualitative (nominal)	Mode	Frequency distribution	Frequency distribution

Question

How does surface affect the size of water drops?

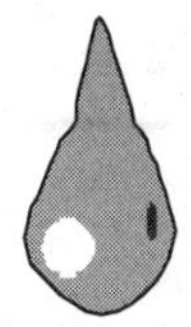

Materials for Each Student

- Small container of water
- Dropping device such as eyedropper, baster, squeeze bottle, straw
- Metric ruler
- Various surfaces: wax paper, aluminum foil, plastic wrap

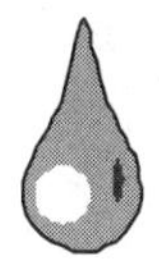

Procedures

1. Fill the dropper with water.
2. Hold the dropper 5 cm above a sheet of wax paper, with the shiny side up; make a drop on the surface.
3. Measure the diameter or longest dimension of the drop to the nearest mm. Record the data.
4. Repeat Steps 2–3 with aluminum foil (shiny side) and plastic wrap (top/shiny side.)

Diameter of drops (mm)

Wax paper ________ **Al foil** ________ **Plastic wrap** ________

5. Record your data in a class data table. Be sure that at least 30 sets of data are entered. If there are fewer than 30 trials, then some students should repeat Steps 1–4 to obtain sufficient trials.

Analyzing Your Data

1. Make stem-and-leaf plots of the data and write a paragraph to summarize findings.
2. Make boxplots. Summarize findings in a data table and paragraph.
3. Calculate the standard deviation/variance. Make an appropriate data table and write a paragraph about results.
4. Write a report about the experiment.

(continued on the following page)

INVESTIGATION 10.2 ▪ Watery Statistics *(continued)*

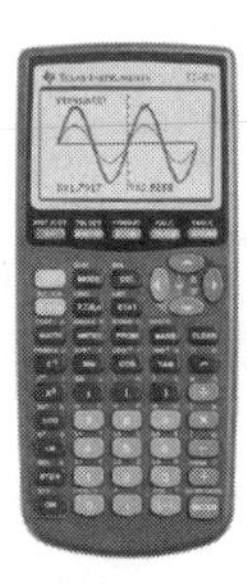

USING TECHNOLOGY

1. In the **STAT** mode of your calculator, enter the water drop measurements in Lists 1, 2, and 3. Sort the measurements in ascending order. Use the ordered data to make ordered stem-and-leaf plots. (See Appendix A, *Using Technology*, for additional help in using the graphing calculator.)
2. In the **STAT** mode select CALC (for calculate) and then 1 VAR (for 1-variable statistics). Depending on the brand of your calculator, you will need to enter the desired list number (e.g., L1), or SET the 1-variable x-list to the desired list number before selecting 1 VAR. *Repeat* the selection process for each set of data by changing the list number. Obtain 1-variable statistics for each list of data including mean, median, quartiles, minimum, maximum, standard deviation, and number. Use this information to make data tables and graphical displays.
3. Set up a different graph for each set of data you want displayed. In setting up the first graph, select boxplot as your graph type and List 1 for the x-list and 1 for the frequency. Repeat to set up the second and third graphs, but enter List 2 or List 3 for the x-list. Graph all three simultaneously.
4. Press **Trace** and use the left and right arrow keys to highlight such values as the minimum, 1st quartile, median, 3rd quartile, and maximum. To make comparisons among the three different boxplots, use the up or down arrows to toggle between graphs.
5. Link your graphing calculator to a computer and download these boxplots to save, enlarge, or print them.

Extending Your Learning

1. What is adhesion? Cohesion? Capillary action?
2. What is the chemical composition of wax paper, aluminum foil and plastic wrap? Use this information to explain differences among the water drops?
3. What other variables could you measure or describe? How could you operationally define them? What type of measures of central tendency and variation/dispersion would be appropriate?
4. What other factors affect the behavior of various liquids on various surfaces? Design experiments to test your hypotheses.

REFERENCES

Glenberg, A. (1996). *Learning from data: An introduction to statistical reasoning* (2nd ed.). Mahwah, NJ: Lawrence Erlbaum Associates.

Landwehr, J.M. & Watkins, A.E. (1994). *Exploring data.* A component of the Quantitative Literacy Series. Palo Alto, CA: Dale Seymour Publications.

Landwehr, J.M., Swift, J., & Watkins, A.E. (1998). *Exploring surveys and information from samples.* A component of the Data-Driven Mathematics Series. Palo Alto, CA: Dale Seymour Publications.

McClave, J.T., Dietrich, Frank, H. II, & Sincich, T. (1997). *Statistics* (7th ed.). Upper Saddle River, NJ: Prentice-Hall, Inc.

Shavelson, R.J. (1996). *Statistical reasoning for the behavioral sciences* (3rd ed.). Boston: Allyn & Bacon.

Yates, D.S., Moore, O.S., McCabe, G.P. (1999). *The Practice of Statistics: TI-83 Graphing Calculator Enhanced.* New York: W.H. Freeman and Company.

Related Web Sites

www.statcan.ca
http://www.mste.uiuc.edu/stat/stat.html
www.lib.umich.edu/libhome/Documents.center/stats.html
www.bls.gov/oreother.htm
http://www.scri.fsu.edu/~dennisl/CMS/sf/sf_details.html (Standard Deviation)

Practice

For the following data set: a) determine appropriate measures of central tendency and dispersion/variation, b) construct stem-and-leaf plots and write a paragraph about findings, c) construct boxplots, data table, and paragraph of results, d) calculate the standard deviation/variance and make appropriate data tables, graphical displays, and paragraphs of results, and e) select the most appropriate method for displaying dispersion/variation and justify your answer.

1. Data on the Effect of Various Percentages of Nitrogen in Fertilizer on the Height of Corn Plants (cm)

Percentage (%) of Nitrogen in Fertilizer												
Trials	**Control (0%)**			**5%**			**10%**			**15%**		
1–3	1.9	0.9	0.8	2.2	2.3	2.4	3.3	2.7	2.5	3.9	3.2	3.1
4–6	1.5	1.5	2.0	2.1	2.6	2.7	3.1	2.8	2.1	3.2	3.0	3.6
7–9	1.0	2.0	1.4	2.9	2.5	2.3	3.0	2.9	2.4	3.7	3.5	3.8
10–12	1.4	0.9	1.3	2.6	2.8	2.6	3.3	3.2	3.3	3.5	3.2	3.2
13–15	0.9	1.4	1.3	2.2	2.6	2.7	3.0	2.9	2.6	3.6	3.9	3.9
16–18	0.8	1.2	1.8	2.5	3.0	2.8	3.0	3.4	2.5	3.7	3.0	3.1
19–21	1.2	1.3	1.3	2.6	2.3	2.6	3.0	3.4	3.1	3.4	4.1	3.5
22–24	0.5	1.4	1.9	2.6	2.8	2.4	2.7	3.2	3.9	3.2	3.4	3.2
25–27	1.5	0.9	1.5	3.1	2.5	2.7	2.7	2.4	3.4	3.8	3.6	3.7
28–30	1.3	1.8	1.6	2.8	2.4	2.5	2.8	3.3	3.5	3.1	3.7	3.5
31–33	1.7	0.8	1.3	2.8	2.9	2.7	3.3	2.7	3.2	3.4	3.9	3.4
34–35	1.0	1.7		3.6	2.8		2.9	3.0		3.5	3.7	

2. Ms. Goldfarb's students thought that sometimes they didn't get their money's worth at a local fast food restaurant when they bought a jumbo order of french fries. Each server seemed to vary a lot in the number of french fries they put in each order. The students designed a study to answer the following question: Does the number of french fries in

(continued on the following page)

Practice *(continued)*

the orders served by an individual server vary? On a particular day they asked people to count the number of french fries they were served. The data the students collected were:

Server 1								Server 2								
32	38	33	32	35	32	34	31	29	31	34	31	31	32	31	32	29
34	34	37	33	36	38	34	37	34	33	30	28	30	29	33	34	32
36	35	36	36	31	33	35		28	29	28	32	34	33	28	28	30
35	31	34	34	33	32	37		33	32	30	29	30	33	34	31	
Server 3								**Server 4**								
32	32	29	30	33	33	33	33	34	29	32	33	36	29	30	30	
32	33	30	33	33	31	29	32	31	34	30	35	32	31	28	29	
33	32	33	32	32	33	33		33	30	34	34	30	35	31		
31	30	31	33	31	32	33		36	35	30	28	34	34	33		

A. Use a leaf-stem plot for each set of the servers' data to determine which if any of the sets of data for a server form a J distribution.

B. Use a leaf-stem plot for each set of the servers' data to determine which if any of the sets of data for a server form a normal distribution.

C. Use a leaf-stem plot for each set of the servers' data to determine which if any of the sets of data for a server form a bimodal or U distribution.

D. Use a leaf-stem plot for each set of the servers' data to determine which if any of the sets of data for a server form a rectangular distribution.

E. What is the range, mean, median, Q_1, Q_3, and standard deviation for each server?

	Server 1	Server 2	Server 3	Server 4
range				
mean				
median (Q_2)				
Q_1				
Q_3				
SD				

F. Which server had the greatest variation in the number of french fries served? How can you tell?

G. Which server had the least variation in the number of french fries served? How can you tell?

CHAPTER 11

Determining Statistical Significance

Objectives

- Determine the level of significance and degrees of freedom for a statistical test and use these concepts to explain the probability of error.
- Identify the appropriate inferential statistic to use for a given set of data.
- Construct an appropriate data table and graph to communicate data.
- Use a checklist to evaluate inferential statistical tests, paragraphs of results, and a conclusion, and to identify needed improvements.

National Standards Connections

- Formulate and revise scientific explanations and models using loci and evidence (NSES).
- Understand sampling and recognize its role in statistical claims (NCTM).
- Test hypotheses using appropriate statistics (NCTM).

By the end of February, Mrs. Smith's biology students completed their experiments and began to organize data into appropriate tables and graphs. Unanswered questions remained. Is the research hypothesis supported by the data? Did the experimental treatment make a difference? What recommendations would you make?

Amy rapidly completed her quantitative data table comparing the height of 75 ornamental mimosa seedlings that were forced to close daily with a control group consisting of an equal number of plants. For three months, the experimental seedlings, age one year, had been stroked at 7:00 a.m. and 7:00 p.m. to force closure. Heights of the experimental and control plants were measured and recorded. Amy knew that the leaf-closing response of plants provided protection and long-term survival; however, she hypothesized that the short-term effects upon growth would be negative because of increased use of energy. Amy reported that stroked plants had a mean height of 80 cm, as compared with 87 cm for the control. Slightly greater variation in plant height occurred within the stroked group (standard deviation = 8.2 cm) than within

the control (standard deviation = 5.8 cm). Amy concluded that the data supported the research hypothesis and recommended that the leaf closing response of houseplants, not typically subject to harsh conditions, be eliminated through selective breeding.

After Ben learned that acetylene, a hydrocarbon gas, hastened the ripening of fruit, he hypothesized that other hydrocarbons produced by the incomplete combustion of fossil fuels would have a similar effect. Ben counted the number of apples displaying various colors when ripened in a smoke-filled atmosphere as compared with a normal (control) atmosphere. Although apples ripened in normal and smoke-filled atmospheres had the same mode, light red, substantial variation occurred in fruit color as shown in Table 11.1 below:

TABLE 11.1 Color of Apples in Various Atmospheres

	Color of Apples				
	Green	Pink	Light Red	Medium Red	Dark Red
Smoke-filled atmosphere (Experimental group)	5	15	35	25	20
Normal atmosphere (Control group)	10	30	45	10	5

Because more ripened fruit occurred in the smoke-filled atmosphere, Ben concluded that the research hypothesis was supported and that fossil fuel pollution could affect fruit ripening.

Based upon descriptive statistics, both Amy and Ben concluded that their research hypotheses were supported and described applications for other groups of mimosa plants and apple trees. However, were the findings reported by Amy and Ben **statistically significant?** What is the probability that the differences occurred by chance and were not a result of the experimental treatment? What inferences can be made from the samples of mimosa plants and apples that could apply to populations of these items? To answer these questions, inferential, rather than descriptive statistics must be used.

Descriptive statistics consist of mathematical procedures that report important characteristics of data, including measures of central tendency and variation. These procedures are described in Chapter 8. **Inferential statistics** expand the researcher's framework from a small group, the sample, to the entire group, the population. Various mathematical procedures, statistical tests, exist for determining the probability that observed differences result from the experimental treatment, rather than from chance. Just as with descriptive statistics, different inferential statistical tests are used with quantitative and qualitative data. Two frequently used tests are the t test for quantitative data and the Chi-square test for qualitative data. The t test is used to determine whether significant differences exist between means; whereas, the Chi-square test evaluates the significance of differences between frequency distributions. Table 11.2 *Appropriate Statistics for Various Data* summarizes the major descriptive and inferential statistics used with quantitative and qualitative data. This chapter describes basic principles of inferential statistics, calculations of the t test and Chi-square test, and reporting of statistical findings.

POPULATIONS AND SAMPLES

A scientist uses samples as a vehicle for investigating a population. The sample, the population from which the sample was drawn, and the target population represent successively larger units of study that are operationally defined by the researcher:

- **Sample:** the specific portion of the population that is selected for study, for example, *the 150 mimosa seedlings used in Amy's study;*
- **Sampled Population:** the population from which the sample was drawn, e.g., *all the mimosa seedlings in the nursery from which Amy obtained her mimosa seedlings;*

TABLE 11.2 Appropriate Statistics for Various Data

Category	Analysis of data	Quantitative data (Continuous)	Qualitative data (Discrete-Categorical)
Descriptive Statistics	Measure of central tendency	Mean	Median Mode
	Measure of variation	Range Variance Standard deviation	Frequency distribution
Inferential Statistics	Statistical test	*t* Test	Chi-Square

- **Target Population:** all units (persons, things, experimental outcomes) of the specific group whose characteristics are being studied, such as *all the mimosa seedlings of the same species.*

The validity of an experiment depends on a precise definition of the population and careful sampling of the defined population. Populations may be very large, such as all white oak trees within the southeastern United States, or very small, such as all students in a given chemistry class. Populations may consist of complete groups (trees, people, motors) or subdivisions within these groups (xylem cells, hearts, or fan belts). For validity, the sampled population must be similar to the target population. If the mimosa seedlings in the nursery Amy used differed greatly from other mimosa seedlings of the same species, the findings would be invalid. Selection of the sample is very important. Random samples in which every individual member of the population has an equal chance of being included are preferred and are an underlying assumption of many statistical tests. Tables of random numbers, generated by computers or located in standard reference texts, may be used in selecting samples. To select a random sample from 2,000 mimosa seedlings in a nursery, Amy would assign a four digit number, starting with 0001, to each plant and then consult a table of random numbers to determine the specific plants to include in the 150-plant sample.

Bias can occur when samples are drawn so that all numbers do not have an equal chance to be chosen. For example, bias would exist if Amy selected all mimosa plants for the experimental group from one part of the greenhouse and plants for the control group from another section. Variations between growth conditions in the two parts of the greenhouse or among the genetic stock of the plants could introduce such bias.

Confidence in experimental findings increases with additional repeated trials. Likewise, confidence in inferences from the sample to the population increases with larger random samples. In a large random sample, a continuous variable such as plant height tends to fall into a bell-shaped frequency distribution, or normal curve, with the sample mean ($\bar{X}$) approximating the population mean (μ). Multiple random samples may be selected from the same population. With descriptive statistics, the mean, variance, and standard deviation of the sample means can be calculated. The means ($\bar{X}$) of the samples tend to fall into a normal curve with the mean of the sample means ($\bar{\bar{X}}$) approximating the population mean (μ). Similarly, the standard deviation of the samples approximates the standard deviation (σ) of the population. The special term, standard error, is used for the standard deviation of the sample means. Relationships among a sample, multiple samples, and the population are depicted in Table 11.3 *Relationships Among Samples of Populations.*

TABLE 11.3 Relationships Among Samples of Populations

Sample (Mimosa plants)	Multiple Samples (100 mimosa plants/sample)	Population (All targeted mimosa plants)
Increased Confidence ↓ 10 plants; 100 plants; 1,000 plants	Increased Confidence ↓ ① ② ③ . . . . ⑩ 10 samples; ① ② ③ ④ ⑤ . . . . 50 50 samples; ① ② ③ ④ ⑤ . . . . 100 100 samples	All mimosa plants
Sample mean = $\bar{X} = \frac{\Sigma X_i}{n}$	Mean of samples = $\bar{\bar{X}}_{samples} = \frac{\Sigma \bar{X}_{samples}}{n_{samples}}$	True mean (μ)
Sample variance = $s^2 = \frac{\Sigma(X_i - \bar{X})^2}{n-1}$	Variance of samples = $s^2_{samples} = \frac{\Sigma\left(\bar{X}_{sample} - \bar{\bar{X}}_{\substack{sample\\means}}\right)^2}{n_{samples} - 1}$	True variance (σ^2)
Sample standard deviation = $s = \sqrt{s^2}$	Standard deviation of samples* = $s_{sample} = \sqrt{s^2_{sample}}$	True standard deviation (σ)
Frequency distribution of plant heights in sample	Frequency distribution of mean plant heights from multiple samples	True frequency distribution of plant heights in population

* The special name, standard error of the mean, is used for the standard deviation of the sample means.

Σ = sum

Practice Set 1

1. Distinguish among the following: (a) sample, sampled population, target population; (b) random sample and biased sample.

2. Anthony investigated the effect of herbicide runoff on the death rate of Daphnia in his grandfather's pond that was located in Grady County, Georgia. Define the sample, sampled population, and target population.

3. Pretend you are Ben and plan to conduct the experiment on the effect of a smoke-filled atmosphere on the ripening of apples. How would you define your sample, sampled population, and target population? How would you obtain your sample?

4. Both Jackie and Tony conducted an experiment on the effect of Brand X paint in preventing the rusting of iron. Jackie used a sample of 10 nails. Tony used a sample of 100 nails. Jackie reported that Brand X paint was ***not*** effective in preventing rusting. Tony reported that Brand X paint was effective. In whose findings would you place the greatest confidence? Why?

HYPOTHESES AND SIGNIFICANCE

Scientists design experiments to determine if a specific research hypothesis or set of hypotheses is supported by the data. A **research hypothesis** often comes from the literature review and suggests the outcome of the experiment. Frequently, research hypotheses contain an if . . . then component that predicts the effect of changing the independent variable on the dependent variable. Examples of research hypotheses include:

- Because of increased energy use, forced closure mimosa plants will be shorter than nonforced closure plants.
- Wood production in trees adjacent to herbicide-treated fields will be less than wood production in trees adjacent to nonherbicide-treated fields.
- If the concentration of Chemical X is increased, then plant growth will be reduced.

After the experiment is completed, the researcher must determine whether differences between the experimental and control groups occurred by chance or reflect true differences. Scientists infer that true differences result from the experimental treatment. With controlled experiments having no other hidden variables, this inference is valid. The larger the difference between groups, the greater the probability that a true difference exists and that the inferred action of the experimental treatment is supported. To determine whether a difference is large enough to support the decision that the experimental treatment made a difference, scientists establish a statistical hypothesis and test it at a specified level of significance.

A **statistical hypothesis** refers to populations and represents the way by which findings are generalized from the sample to the population. As previously mentioned, multiple samples can be selected from a population. Because the means of these samples form a normal distribution, the means of two samples from the same population may not be identical. If samples are drawn from two different populations, one would expect the two sample means to be different. The statistician's challenge is to determine whether the means of two samples are sufficiently different to support the decision that a true difference exists and that two different populations are represented. The statistician begins by assuming that the two samples represent the same population and have identical means. Therefore, any observed difference between two sample means occurred by chance and is not significant. These assumptions are expressed through the **null hypothesis.** Several acceptable formats exist for expressing the null hypothesis in words. One option is to state that the two population means are equal; a second is to state that they are not significantly different. When the null hypothesis is expressed in mathematical symbols, the means of the two populations (1 and 2) are written as equal. Examples of null hypotheses follow.

- The mean height of forced closure mimosa plants is not significantly different from the mean height of nonforced closure mimosa plants.

$$\mu_{\text{Forced}} = \mu_{\text{Non-Forced}}$$

- Mean wood production in trees adjacent to herbicide and nonherbicide treated fields is not significantly different.

$$\mu_{\text{Herbicide}} = \mu_{\text{Non-Herbicide}}$$

- The mean heights of plants exposed to 0 percent, 10 percent, 20 percent, and 30 percent Chemical X are equal.

$$\mu_{0\%} = \mu_{10\%} = \mu_{20\%} = \mu_{30\%}$$

LEVEL OF SIGNIFICANCE

The level of significance (α) required for statistical significance is determined by the researcher and is affected by sample size and the nature of the experiment. Common levels of significance (α) are 0.05, 0.01, and 0.001. The level of significance (for example, 0.05) communicates the probability that the researcher made a mistake in rejecting the null hypothesis. At 0.05, the probability of error in rejecting the null hypothesis is 5/100; whereas, at 0.01 the probability is 1/100. When an error would have a major impact, such as in drug testing, the level of significance is reduced to 0.001 or 1/1000.

Because the means of multiple samples from a population follow a normal distribution, the probability of obtaining a specific difference between the means of two samples of a given size can be determined. A section from a *t* sampling distribution is presented in Table 11.4 below.

TABLE 11.4 Excerpt from a *t* Sampling Distribution

	Probability (Level of Significance)			
df	*0.1*	*0.05*	*0.01*	*0.001*
5	2.015	2.571	4.032	6.859
.				
.				
10	1.812	2.228	3.169	4.587
.				
.				
20	1.725	2.086	2.845	3.850
.				
.				
40	1.684	2.025	2.704	3.551
.				
.				
	1.645	1.960	2.576	3.291

The influence of sample size is reflected in the degrees of freedom (df) located on the left side of the section from the *t* sampling distribution. Degrees of freedom represent the number of independent observations in a sample. In a sample of six plants, the mean plant height would be determined by adding the six values for plant height and dividing by the number of plants.

$$\bar{X} = \frac{50 + 60 + 30 + 50 + 50 + 60}{6} = \frac{300}{6} = 50$$

Similarly, if the mean of the six plants and the individual heights of five plants were known, the individual height of the sixth plant could be determined. The sixth value is not free to vary but is determined by the first five; therefore, only five degrees of freedom exist.

$$50 = \frac{50 + 60 + 30 + 50 + 50 + x_i}{6}$$

In a sample of n numbers with a fixed mean, the degrees of freedom are equal to n – 1. In Amy's experiment, both the experimental and control groups contained 75 mimosa plants; thus, the total degrees of freedom is 148, (75 – 1) for the experimental group plus (75 – 1) for the control group.

The influence of sample size on the expected difference between means (statistical value) can be seen from examining numerical relationships in the vertical columns of the section from the *t* sampling distribution. The larger the sample (degrees of freedom), the smaller the difference between means (statistical value) required to support the decision that the means are not from the same population. Remember, scientists have more confidence in experiments with larger numbers of subjects and repeated trials. At degrees of freedom of 5, 10, 20, and 40, statistical values required for significance are 2.571, 2.228, 2.086, and 2.025, respectively, at the 0.05 level.

The influence of the level of significance (α) can be seen from examining the numerical relationships in a horizontal row of the same section of a *t* sampling distribution. The smaller the level of significance and error rate, the larger the difference between means (statistical value) required for significance. For example, at 20 degrees of freedom, *t* values are 2.086, 2.845, and 3.850, respectively, for the 0.05, 0.01, and 0.001 levels of significance.

To make an appropriate decision regarding the null hypothesis, statisticians compare a calculated value with the table value at the appropriate degrees of freedom and level of significance (see Table 11.5 *Sample Distribution for t test* (for a *t* sampling distribution). If the calculated value is smaller than the required table value, the **null hypothesis is not rejected.** When the calculated value equals or exceeds the required table value, the **null hypothesis is rejected.** For example, if a table value of 2.600 was required for significance at the 0.05 level, the following decisions would be appropriate:

Calculated value	Table value (0.5 level)	Decision about null hypothesis	Probability (p)
2.542	2.600	Not Reject	>0.05
2.600	2.600	Reject	0.05
2.785	2.600	Reject	<0.05

PROBABILITY OF ERROR

When the calculated and the table values are equal, the probability (p) of error in rejecting the null hypothesis is equal to the level of significance (α). When the calculated value is larger, the probability (p) of error is less than the level of significance (α). With detailed tables, the exact probability level can be determined by locating the probability level at which the calculated value equals the table value, for example, $p = 0.001$ rather than $p < 0.05$.

Rejection of the null hypothesis supports the alternative decision that a true difference exists between the means and that the samples represent different populations. Because a researcher generally hypothesizes a difference between means, rejection of the null hypothesis lends support to the research hypothesis. In the rare instances where a researcher hypothesizes equivalent means, nonrejection of the null hypothesis lends support to the research hypothesis.

THE *t* TEST

For quantitative data, the *t* test can be used to determine if observed differences between means of two groups are statistically significant. The *t* test is essentially a ratio that compares the difference between two means with the total standard deviation within the groups. The resulting quotient is the number of standard errors of the mean (standard deviation) between the two populations.

Several formulas exist for calculating *t.* The specific formula used depends on the number of measurements in each sample and the relationship between the measurement in the two groups being compared. When no relationship exists between measurements in the two groups, for example, experimental and control groups, an **uncorrelated *t* test** is used. Two basic formulas for calculating an uncorrelated *t* test are provided below:

Formula 1: Uncorrelated *t* Test with Equal Sample Size

$$t = \frac{\bar{X}_1 - \bar{X}_2}{\sqrt{\frac{s_1^2 + s_2^2}{n}}}$$

$\bar{X}_1$ = mean of Group 1
$\bar{X}_2$ = mean of Group 2
s_1^2 = variance of Group 1
s_2^2 = variance of Group 2
n = number of items or measurements in Sample 1 = Sample 2

Formula 2: Uncorrelated *t* Test with Unequal Sample Size

$$t = \frac{\overline{X}_1 - \overline{X}_2}{\sqrt{\frac{(n_1 - 1)s_1^2 + (n_2 - 1)s_2^2}{n_1 + n_2 - 2} \cdot \left(\frac{1}{n_1} + \frac{1}{n_2}\right)}}$$

$\overline{X}_1$ = mean of Group 1
$\overline{X}_2$ = mean of Group 2
s_1^2 = variance of Group 1
s_2^2 = variance of Group 2
n_1 = number of items or measurements in Sample 1
n_2 = number of items or measurements in Sample 2

When measurements in the groups are related, such as pretest and posttest data, a correlated *t* test is used. For use of the correlated *t* test and other applications, such as the significance of a difference between percents and one- and two-tailed *t* test, consult the statistical references listed at the end of the chapter.

TABLE 11.5 Sampling Distribution for t Test

Degrees of freedom	Probability (Level of significance)			
	0.1	0.05	0.01	0.001
1	6.314	12.706	63.657	636.619
2	2.920	4.303	9.925	31.598
3	2.353	3.182	5.841	12.924
4	2.132	2.776	4.604	8.610
5	2.015	2.571	4.032	6.864
6	1.943	2.447	3.707	5.959
7	1.895	2.365	3.499	5.408
8	1.860	2.306	3.355	5.041
9	1.833	2.262	3.250	4.781
10	1.812	2.228	3.169	4.587
11	1.796	2.201	3.106	4.437
12	1.782	2.179	3.055	4.318
13	1.771	2.160	3.012	4.221
14	1.761	2.145	2.977	4.140
15	1.753	2.131	2.947	4.073
16	1.746	2.120	2.921	4.015
17	1.740	2.110	2.898	3.965
18	1.734	2.101	2.878	3.922
19	1.729	2.093	2.861	3.883
20	1.725	2.086	2.845	3.850
21	1.721	2.080	2.831	3.819
22	1.717	2.074	2.819	3.792
23	1.714	2.069	2.807	3.767
24	1.711	2.064	2.797	3.745
25	1.708	2.060	2.787	3.725
26	1.706	2.056	2.779	3.707
27	1.703	2.052	2.771	3.690
28	1.701	2.048	2.763	3.674
29	1.699	2.045	2.756	3.659
30	1.697	2.042	2.750	3.646
40	1.684	2.021	2.704	3.551
60	1.671	2.000	2.660	3.460
120	1.658	1.980	2.617	3.373
∞	1.645	1.960	2.576	3.291

Practice Set 2

1. Distinguish among the following:

   A. null hypothesis, research hypothesis
   B. degrees of freedom, sample size, level of significance
   C. calculated statistical value, sampling distribution of a statistic

2. For the following levels of significance, indicate the probability that differences occurred by chance and the probability that differences resulted from the experimental treatment: a) 0.05, b) 0.03, c) 0.15, d) 0.40, e) 0.001.

3. Write a null hypothesis for each of the following research hypotheses:

   A. If the amount of Chemical X added to water is increased (0, 1, 2, 3 scoops), then the mean temperature of the solution will increase;
   B. Because of higher concentration of automobile pollution, loblolly pines along an interstate will exhibit a lower mean height than trees along rural roads;
   C. Lower stress levels, as exhibited by blood pressure, will occur in people who exercise;
   D. If the slope of a stream is increased (10°, 20°, 30°), then the mean sediment load will increase.

4. For each of the following degrees of freedom and levels of significance, find the statistical value required for significance (Note: Use the sampling distribution of *t* located on page 137): (a) df $= 7$, $\alpha = 0.05$; (b) df $= 18$, $\alpha = 0.01$; (c) df $= 30$, $\alpha = 0.001$; (d) df $= 60$, $\alpha = 0.05$.

5. Several students conducted statistical tests on their experimental data. Below are the calculated statistical values and the sampling distribution values required for significance at the 0.05 level. Indicate which decision should be made by the researcher about the (a) null hypothesis—reject or not reject—and (b) research hypothesis—support or not support.

   A. Larry's calculated value = 6.954; sampling distribution value = 6.954 at 0.01 level of significance; Larry hypothesized differences.
   B. Gloria's calculated value = 8.254; at $\alpha = 0.05$, df $= 27$, sampling distribution value = 7.954; Gloria hypothesized differences.
   C. Janet's calculated value = 10.784; sampling distribution value = 13.724 at 0.05 level of significance; Janet hypothesized no differences.

6. Using 0.05 as the maximum level required for significance, indicate what level of significance you would establish for each of the following experiments. Justify your answer.

   A. The Effect of Three Coatings on the Rusting of Iron; sample size = 100.
   B. The Effectiveness of Compound Z in Preventing Hardening of the Arteries, sample size = 500.
   C. The Rate of Molding of Bread in Plastic, Glass, Metal, and Wooden Containers; sample size = 15.

TWO TREATMENT GROUPS

Scenario: John read that farmers in Japan routinely subject plants to stress before transplanting from the greenhouse to the field. Methods of causing stress included pulling on the plants and hitting them with straw rakes. John decided to investigate this technique by growing two groups of bean plants (10/group) in a greenhouse for 15 days during which time the plants in one group were pulled on three times daily at 8:00 in the morning and at 4:00 in the afternoon. The plants were then transplanted to a field. John hypothesized that stressed plants would exhibit greater mean height after transplanting than the nonstressed plants (control). Plant heights after 30 days are given below.

Stressed Plants	*Nonstressed Plants*
55.0	48.0
65.0	65.0
50.0	59.0
57.0	57.0
59.0	51.0
73.0	63.0
57.0	65.0
54.0	58.0
62.0	44.0
68.0	50.0

Step 1: State the null hypothesis.

Null Hypothesis: The mean height of stressed plants is **not significantly different** from the mean height of nonstressed plants.

H_0: $\mu_{stressed} = \mu_{nonstressed}$

Step 2: Establish the level of significance.

Alpha (α) = 0.05

Step 3: Calculate the means.

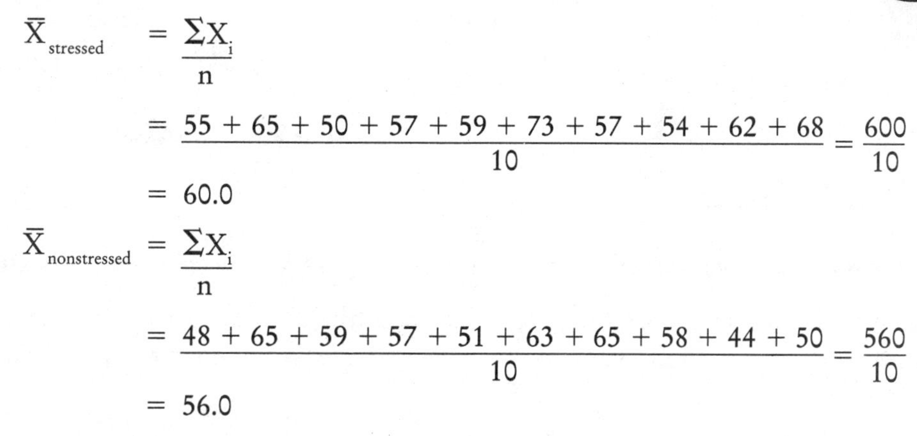

$$\bar{X}_{stressed} = \frac{\Sigma X_i}{n}$$

$$= \frac{55 + 65 + 50 + 57 + 59 + 73 + 57 + 54 + 62 + 68}{10} = \frac{600}{10}$$

$$= 60.0$$

$$\bar{X}_{nonstressed} = \frac{\Sigma X_i}{n}$$

$$= \frac{48 + 65 + 59 + 57 + 51 + 63 + 65 + 58 + 44 + 50}{10} = \frac{560}{10}$$

$$= 56.0$$

Step 4: Calculate the variance.

$$s^2_{\text{stressed}} = \frac{\Sigma(X_i - \bar{X})^2}{n - 1}$$

$$= \frac{(55 - 60)^2 + (65 - 60)^2 + (50 - 60)^2 \ldots + (68 - 60)^2}{10 - 1}$$

$$= 49.1$$

$$s^2_{\text{nonstressed}} = \frac{\Sigma(X_i - \bar{X})^2}{n - 1}$$

$$= \frac{(48 - 56)^2 + (65 - 56)^2 + (59 - 56)^2 \ldots + (50 - 56)^2}{10 - 1}$$

$$= 54.9$$

Step 5: Calculate t.

$$t = \frac{\bar{X}_1 - \bar{X}_2}{\sqrt{\frac{s_1^2 + s_2^2}{n}}}$$

$$t = \frac{60 - 56}{\sqrt{\frac{49.1 + 54.9}{10}}} = \frac{4}{\sqrt{\frac{104.0}{10}}} = \frac{4}{\sqrt{10.4}} = \frac{4}{3.2}$$

$$t = 1.25 \sim 1.3$$

Step 6: Determine the degrees of freedom.

df = (Number of stressed plants – 1) + (Number of nonstressed plants – 1)

df = (10 – 1) + (10 – 1)

df = 9 + 9 = 18

Step 7: Determine the significance of the calculated t.

At df = 18, α of 0.05, $t = 2.101$; the calculated t of $1.3 < 2.101$ and **not significant** at the 0.05 level ($p > 0.10$).

Step 8: Decide to reject or not reject the null hypothesis.

Because the calculated value of t is **not significant,** the null hypothesis is **not rejected.**

H_0: $\mu_{\text{stressed}} = \mu_{\text{nonstressed}}$

Step 9: Determine whether the statistical findings support the research hypothesis.

Because the null hypothesis was **not rejected** at the 0.05 level of significance, the research hypothesis that stressed plants would have a greater mean height than nonstressed plants was **not supported.**

Step 10: Construct a data table that communicates both descriptive and inferential statistics.

In Chapter 8, a model for constructing a quantitative data table that included descriptive information on the sample (mean, variance, and standard deviation) was described. Simply expand the table to include the results of the test as shown in Table 11.6 *Effect of Stress on the Mean Height of Bean Plants*.

TABLE 11.6 Effect of Stress on the Mean Height of Bean Plants

Descriptive information	Stressed group	Nonstressed group
Mean	60.0	56.0
Variance	49.1	54.9
Standard deviation	7.0	7.4
1 SD (68% Band)	53.0 – 67.0	48.6 – 63.4
2 SD (95% Band)	46.0 – 74.0	41.2 – 70.8
3 SD (99% Band)	39.0 – 81.0	33.8 – 78.2
Number	10	10
Results of t test	$t = 1.3$ df = 18 t of $1.3 < 2.101$	$\alpha = 0.05$ $p > 0.10$

Step 11: Write a paragraph describing results.

In Chapter 9 you learned a structured process for writing about data. Using a similar set of steps you could write sentences that communicate critical information.

Steps	*Example*
1. Write a topic sentence stating the independent and dependent variables, and a reference to tables or graphs.	Effects of stress on the height of bean plants are summarized in Table 11.6.
2. Write sentences comparing the measures of central tendency (means) and variation (standard deviations) of the groups.	Stressed plants showed a greater mean height (60.0 cm) than nonstressed plants (56.0 cm). Variations within the groups were similar, with stressed plants having a standard deviation of 7.0 and nonstressed plants a standard deviation of 7.4. Ninety-five percent of the stressed plants fell within the range of 46.0 to 74.0 cm, as opposed to nonstressed plants, which ranged from 41.2 to 70.8 cm.
3. Write sentences describing the statistical test, level of significance, and null hypothesis.	The uncorrelated t test was used to test the following null hypothesis at the 0.05 level of significance: The mean height of stressed plants is not significantly different from the mean height of nonstressed plants.

4. Write sentences comparing the calculated value with the required statistical values and make a statement about rejection of the null hypothesis.

   The null hypothesis was not rejected ($t = 1.3 < 2.101$ at df = 18; $p > 0.10$).

5. Write sentences stating support of the research hypothesis by the data.

   The data did not support the research hypothesis that stressed plants would have a different mean height after planting than nonstressed plants.

Step 12: Write an appropriate conclusion.

In Chapter 9 you learned a structure process for writing conclusions. Using the following questions as a guideline, you could write a concluding paragraph that includes the results of the statistical test.

Questions	*Example*
1. What was the purpose of the experiment?	The effect of stress on the growth of bean plants was investigated by comparing the growth of bean plants subjected to stress for 15 days with a control (nonstressed plants).
2. What were the major findings? (Focus on results of the statistical test).	No significant difference existed between the mean height of stressed plants and nonstressed plants 30 days after transplanting.
3. Was the research hypothesis supported by the data?	The research hypothesis that stressed plants would have a different mean height was not supported.
4. How did your findings compare with those of other researchers?	In contrast, Japanese farmers found that hitting and pulling rice plants were beneficial.
5. What possible explanations can you offer for your findings?	Possible explanations include differences in the methods of administering stress or the type of plant, for example, monocots (rice) versus dicots (beans).
6. What recommendations do you have for further study and for improving the experiment?	Additional investigations using various sources of stress at more frequent intervals with both monocots and dicots should be conducted. Improved experimental design techniques should be implemented, including a larger sample size and a longer growing period.

THREE TREATMENT GROUPS

Because a t test allows only for the comparison of two means, multiple t tests would be necessary if more than two experimental groups were involved. For example, if an experiment involved three treatment groups (A,B,C), the number of possible t tests would be three (A = B; A = C; B = C). With four and five treatment groups, the number of t tests would increase to 6 and 10, respectively. The probability of obtaining a significant t value by chance increases with the number of t tests conducted. Additionally, interactions exist among groups. For these reasons, multiple t tests are not recommended if options exist for using more sophisticated statistical tests such as analysis of variance. Many judges of competitions are lenient because they realize that the concept of a statistical test is of primary importance, rather than the sophistication of the test. However, problems with multiple t tests should be recognized and appropriate actions taken to minimize obtaining significant differences by chance. These include changing the level of significance (from 0.05 to 0.01) and conducting the minimum number of t tests necessary to determine support for the research hypotheses. For example, if you hypothesized that 10% and 20% Chemical X would promote plant growth while 30% X would retard growth, as compared with a control, only 3 t tests, not 6, would be required.

Required for Testing Hypotheses	Maximum That Could Be Conducted
10% = 0%	10% = 0%
20% = 0%	10% = 20%
30% = 0%	10% = 30%
	20% = 0%
	20% = 30%
	30% = 0%

The following experiment illustrates a typical use of multiple t tests by high school students like you.

Scenario: Laura read about recycling of plant materials (grass and leaves) and their use as fertilizer by making a compost pile. Laura decided to investigate the relative effectiveness of commercial fertilizer and compost in promoting the growth of radish plants. Laura hypothesized that plants grown with compost and plants grown with commercial fertilizer would exhibit similar heights and that both fertilized groups would exhibit greater heights than the control. Plant heights (cm) after three weeks of growth are given below.

Commercial Fertilizer	Compost Fertilizer	No Fertilizer (Control)
5.0	5.0	3.0
6.0	6.0	5.0
10.0	4.0	2.0
6.0	7.0	3.0
8.0	3.0	4.0
6.0	3.0	6.0
5.0	4.0	6.0
8.0	7.0	3.0
6.0	6.0	5.0
10.0	5.0	3.0

Step 1: State the null hypothesis.

$1H_0$: $\mu_{compost} = \mu_{commercial}$
$2H_0$: $\mu_{compost} = \mu_{control}$
$3H_0$: $\mu_{commercial} = \mu_{control}$

Step 2: Establish the level of significance.

Alpha (α) = 0.01

Step 3: Calculate the means.

$$\bar{X}_{commercial} = \frac{5 + 6 + 10 \ldots. + 10}{10} = 7$$

$$\bar{X}_{compost} = \frac{5 + 6 + 4 \ldots. + 5}{10} = 5$$

$$\bar{X}_{control} = \frac{3 + 5 + 2 \ldots. + 3}{10} = 4$$

Step 4: Calculate the variance.

$$s^2_{commercial} = \frac{\Sigma(X_i - \bar{X})^2}{n - 1}$$

$$s^2_{commercial} = \frac{(5-7)^2 + (6-7)^2 \ldots. (10-7)^2}{10-1}$$

$$= 3.6$$

$$s^2_{compost} = \frac{(5-5)^2 + (6-5)^2 + (4-6)^2 + \ldots. (5-6)^2}{10-1}$$

$$= 2.2$$

$$s^2_{control} = \frac{(3-4)^2 + (5-4)^2 + \ldots. (5-4)^2 + (3-4)^2}{10-1}$$

$$= 2.0$$

Step 5: Calculate <u>t</u>.

$$t = \frac{\bar{X}_1 - \bar{X}_2}{\sqrt{\frac{s_1^2 + s_2^2}{n}}}$$

$$t_{commercial\ vs.\ compost} = \frac{7.0 - 5.0}{\sqrt{\frac{3.6 + 2.2}{10}}} = 2.62 \sim 2.6$$

$$t_{compost\ vs.\ control} = \frac{5.0 - 4.0}{\sqrt{\frac{2.2 + 2.0}{10}}} = 1.51 \sim 1.5$$

$$t_{commercial\ vs.\ control} = \frac{7.0 - 4.0}{\sqrt{\frac{3.6 + 2.0}{10}}} = 4.01 \sim 4.0$$

Step 6: Determine the degrees of freedom.

df = (Number of plants in commercial – 1) + (Number of plants in compost – 1);
df = 9 + 9 = 18.

Similarly, the degrees of freedom are 18 for the other two t tests.

Step 7: Determine the significance of the calculated t.

At df = 18, α of 0.01, $t = 2.878$; as outlined below, only the calculated t for the commercial-control comparison is significant at the 0.01 level.

$t_{\text{commercial vs. compost}} = 2.6 < 2.878$; $0.01 < p < 0.05$ not significant
$t_{\text{compost vs. control}} = 1.5 < 2.878$; $p > 0.01$ not significant
$t_{\text{commercial vs. control}} = 4.0 > 2.878$; $p < 0.001$ significant

Step 8: Decide whether to reject the null hypotheses.

Because the calculated values of t are not significant for the commercial-compost and compost-control comparisons, the null hypotheses are **not rejected.** For the commercial-control comparison, the calculated t is significant; thus, the null hypothesis is **rejected.**

Step 9: Determine if the statistical findings support/do not support the research hypothesis.

Results of the uncorrelated t test support the first research hypothesis that plants grown with commercial and compost fertilizer would have similar heights. The hypothesized greater growth of plants fertilized with commercial fertilizer, as compared with a control, was also supported by the data. Because no significant difference was found between the height of plants fertilized with compost and the control, the research hypothesis of increased growth with compost was not supported.

Step 10: Construct a data table (see Table 11.7 *Effect of Fertilizer on the Mean Height (cm) of Bean Plants*).

TABLE 11.7 Effect of Fertilizer on the Mean Height (cm) of Bean Plants

Descriptive information	Commercial	Compost	Control
Mean	7.0	5.0	4.0
Variance	3.6	2.2	2.0
Standard deviation	1.9	1.5	1.4
1 SD (68% Band)	5.1 – 8.9	3.5 – 6.5	2.6 – 5.4
2 SD (95% Band)	3.2 – 10.8	2.0 – 8.0	1.2 – 6.8
3 SD (99% Band)	1.3 – 12.7	0.5 – 9.5	0 – 8.2
Number	10	10	10

Results of *t* test

Commercial vs Compost	$t = 2.6$; $.01 < p < 0.05$
Compost vs Control	$t = 1.5$; $p > 0.01$
Commercial vs Control	$t = 4.0$; $p < 0.001$

At df of 18; α of 0.01; $t = 2.878$ for significance

Step 11: Write a paragraph describing results.

The relative effects of commercial fertilizer and compost on the growth of bean plants, as compared with a control group, are summarized in Table 11.7. The mean height (7.0 cm) of plants grown with commercial fertilizer was greater than the mean height of plants grown with compost (5.0 cm) or the control (4.0 cm). Greater variation in plant height was found in the commercial fertilizer group, standard deviation of 1.9, than in the compost group (SD = 1.5) or the control (SD = 1.4). The uncorrelated t test was used to test three null hypotheses at the 0.01 level of significance.

Null Hypothesis 1: The mean height of plants fertilized with compost is not significantly different from the mean height of plants fertilized with commercial fertilizer.

Null Hypothesis 2: The mean height of plants fertilized with compost is not significantly different from the mean height of plants receiving no fertilizer (control).

Null Hypothesis 3: The mean height of plants fertilized with commercial fertilizer is not significantly different from the mean height of plants receiving no fertilizer (control).

Null Hypothesis 1 was not rejected; the mean height of plants fertilized with commercial fertilizer was not significantly greater than the mean height of plants fertilized with compost ($t = 2.62 < 2.878$ at df = 18; $0.01 < p < 0.05$). Null Hypothesis 2 was not rejected, because the mean heights of plants receiving compost and the control were equivalent ($t = 1.51 < 2.878$ at df = 18; $p > 0.01$). Null Hypothesis 3 was rejected, with plants receiving commercial fertilizer exhibiting greater mean height than the control group ($t = 4.01 > 2.878$ at df = 18; $p < 0.01$). These findings support the research hypotheses that commercial fertilizer and compost would have similar effects on plant growth and that commercial fertilizer would be superior to the control. The data do not support the hypothesis that compost would promote plant growth compared with a control.

Step 12: Write an appropriate conclusion.

The relative effectiveness of commercial fertilizer and compost in promoting radish growth was investigated. Mean height of plants receiving commercial fertilizer was significantly greater than the control. No significant differences in heights were found between plants receiving compost and commercial fertilizer. The findings support the hypothesized equivalency of commercial and compost fertilizer and the superiority of commercial fertilizer, as compared with a control. The hypothesized superiority of compost as compared with a control was not supported by the data. Black (1999), Stone (1997), and Jones (1998) reported superior results with compost. Age of compost in these experiments differed and a longer growing period (45–60 days) was used. Further experimentation is necessary to determine if other types of compost of different ages are beneficial to plant growth over an extended period.

Practice Set 3

1. Gail Adams investigated the effect of time of application of an herbicide on the yield of soybeans. Gail hypothesized that herbicides applied early in the morning would be more effective than at midday. Determine if Gail's hypothesis was supported.

	8:00 A.M.	12:00 P.M.
Mean Yield *(g)*	70.8	77.4
Standard deviation	30.8	31.2
Number *(n)*	30	30

2. Using Amy's data from the introduction to this chapter, determine whether the mean height of the forced-closure mimosa seedlings differed significantly from the control group.

3. Under the supervision of a mentor, David Setchel investigated the effect of testosterone and estrogen on the prostate weight (mg/100 g body mass) of rats. Descriptive information on the samples was provided in his paper:

	Testosterone	Estrogen	Control
Mean	70.34	53.55	53.16
Variance	199.34	190.35	57.27
Standard deviation	14.14	13.80	7.57
Number *(n)*	20	20	20

Determine if David's research hypothesis that prostate mass would be positively affected by testosterone and negatively affected by estrogen was supported by the data.

(continued on the following page)

Practice Set 3 *(continued)*

4. In the practice problems in Chapter 10, Ms. Goldfarb's students thought that sometimes they didn't get their money's worth at a local fast food restaurant when they bought a jumbo order of french fries. Different servers seemed to vary a lot in the number of french fries they put in each order. The students designed a study to see if this was true, and on a particular day, they asked people to count the number of french fries they were served.

   A. Use a *t* test to determine if there is a statistically significant difference in the numbers of french fries each server served.

   B. Between which servers do you think there is a statistically significant difference?

Servers / Test	*t* value	p value	Significant Y/N?
1 & 2			
1 & 3			
1 & 4			
2 & 3			
2 & 4			
3 & 4			

5. Conduct Investigation 11.1, *Magnetic Time.* Use a *t* test to analyze your data. Communicate your findings through an appropriate data table, graph, and paragraph.

INVESTIGATION 11.1 ▪ Magnetic Time

Question

How does separation method affect time to separate a mixture?

Mixture	Magnetic

Materials for Each Group of Two Students

- Magnet
- White paper, letter size (8.5 in. x 11.0 in.)
- Plastic zip bag (pint size) with mixture of magnetic and non-magnetic objects. Examples include paper clips and pinto beans or metal paper clips and non-magnetic (brass) paper fasteners or plastic and metal paper clips.
- Watch or clock with second hand.

Procedure

1. Draw a line to divide a sheet of notebook sized paper in half. Label one side "mixture" and the other side "magnetic."
2. Pour the contents of the zip bag on the side labeled "mixture." Be sure all of the items are on that side.
3. By hand, sort the magnetic objects into the other side labeled "magnetic." Record the time (sec).
4. Return the objects to the zip bag, close, and shake to evenly distribute items.
5. Repeat Steps 2–4 using a magnet to separate the items. Record the time (sec).
6. Have your partner do Steps 1–5.
7. Record your data and your partner's in a class data table. Be sure that at least 30 sets of data are entered. If there are fewer than 30 trials, then some students should repeat Steps 1–6 to obtain sufficient trials.

Analyzing Your Data

1. Make a data table to display the class data. Use Table 11.4 and/or 11.5 as a model.
2. Conduct a *t* test to determine if there is a significant difference in the time required for sorting the objects by hand and with a magnet. Enter the appropriate information into the data table.
3. Write a paragraph of results. (Hint: See paragraph accompanying Table 11.4 and/or 11.5.)

(continued on the following page)

INVESTIGATION 11.1 ▪ Magnetic Time *(continued)*

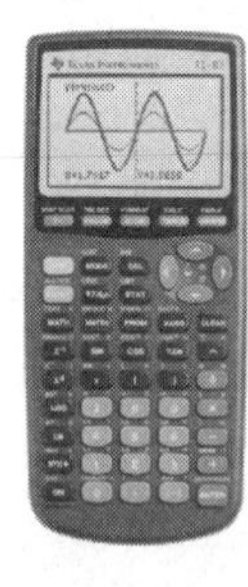

USING TECHNOLOGY

1. In the **STAT** mode of your calculator, enter the time to separate by hand in List 1 and the time to separate with a magnet in List 2. (See Appendix A, *Using Technology*, for additional help in using the graphing calculator.)
2. Obtain 1-variable statistics for each set of data, such as the mean, standard deviation, minimum, maximum, and number. In the **STAT** mode select CALC (for calculate) and then 1 VAR (for 1-variable statistics). Depending on the brand of your calculator, you will need to enter the desired list number (e.g., L1), or SET the 1-variable x-list to the desired list number before selecting 1 VAR. *Repeat* the selection process for the second set of data by changing the list number.
3. Then select **TESTS** and choose the 2-sample *t* test.
4. Enter List 1 and List 2 as the locations of the data, select *Yes* for pooled data, and then calculate.
5. Among the values provided, are the *t* value, the probability the differences are due to chance, the degrees of freedom, and the means for both data sets.

Extending Your Learning

1. How are magnets used in industry, manufacturing, and mining to separate objects?
2. What affects the strength of a magnet? Do you think you would get the same results with magnets of different sizes, ages, compositions? How could you design experiments to test your hypothesis? What other ways could you measure the dependent variable?
3. Do you think that the type of mixture affects the task? What other mixtures could you use? How could you design experiments to test your hypothesis?
4. What other factors might affect the outcome of this task?

CHI-SQUARE

For qualitative data, chi-square (χ^2) can be used to determine if differences between frequency distributions are statistically significant. An observed frequency distribution may be compared with an expected or theoretical frequency distribution for goodness of fit using the following formula.

$$\chi^2 = \sum \frac{(O - E)^2}{E}$$

χ^2 = Chi-square
Σ = Sum of the Values
O = Observed Frequency Distribution
E = Expected Frequency Distribution

The calculated χ^2 value is compared with a sampling distribution of χ^2 to determine significance (see Table 11.8 *Chi-Square Sampling Distribution*). Chi-square may also be used to evaluate whether two variables are associated or related; for use of such contingency tables, consult the statistical references listed at the end of the chapter.

GOODNESS OF FIT

Scenario: Mary read that bees were attracted to the color yellow as opposed to red, blue, or white. She wondered if crickets would show a color preference. To test her hypothesis that crickets would be differentially attracted to colors, she placed 100 crickets in a container. The bottom of the container was divided into four equal sections covered by red, blue, yellow, or white paper. She observed the number of crickets on each color paper one hour after placing them in the container. The distribution of crickets was: 30 red, 40 blue, 12 yellow, and 18 white. By chance alone, an equal number of crickets on each color of paper would be expected.

Step 1 State the null hypothesis.

Null Hypothesis: The frequency distribution of crickets on various colors is **not significantly different** from the frequency distribution predicted by chance.

Observed Frequency Distribution on Various Colors = Expected (Chance) Frequency Distribution on Various Colors

Step 2 Establish the level of significance.

Alpha (α) = 0.05

Step 3 Determine the observed frequency distribution.

		Red	*Blue*	*Yellow*	*White*
Overall Distribution	=	30	40	12	18

Step 4 Determine the expected theoretical frequency distribution.

By chance, one would expect an equal distribution of crickets across the four colors. Because a total of 100 crickets was used, 1/4 of 100 or 25 crickets are expected per color.

		Red	*Blue*	*Yellow*	*White*
Expected Distribution	=	25	25	25	25

TABLE 11.8 Chi-Square Sampling Distribution

Degrees of Freedom	Probability (Level of significance)			
	0.1	0.05	0.01	0.001
1	2.706	3.841	6.635	7.879
2	4.605	5.991	9.210	10.597
3	6.251	7.815	11.345	12.830
4	7.779	9.488	13.277	14.860
5	9.236	11.070	15.086	16.750
6	10.645	12.592	16.812	18.548
7	12.017	14.067	18.475	20.278
8	13.362	15.507	20.090	21.955
9	14.684	16.919	21.666	23.589
10	15.987	18.307	23.209	25.188
11	17.275	19.675	24.725	26.757
12	18.549	21.026	26.217	28.300
13	19.812	22.362	27.688	29.819
14	21.064	23.685	29.141	31.319
15	22.309	24.996	30.578	32.801
16	23.542	26.296	32.000	34.267
17	24.769	27.587	33.409	35.718
18	25.989	28.869	34.805	37.156
19	27.204	30.144	36.191	38.582
20	28.412	31.410	37.566	39.997
21	29.615	32.670	38.932	41.401
22	30.813	33.924	40.289	42.796
23	32.007	35.172	41.638	44.181
24	33.196	36.415	42.980	45.558
25	34.382	37.652	44.314	46.928
26	35.563	38.885	45.642	48.290
27	36.741	40.113	46.963	49.645
28	37.916	41.337	48.278	50.993
29	39.088	42.559	49.588	52.336
30	40.256	43.773	50.892	53.672

Step 5 Calculate χ^2.

$$\chi^2 = \frac{(30-25)^2}{25} + \frac{(40-25)^2}{25} + \frac{(12-25)^2}{25} + \frac{(18-25)^2}{25}$$

$$= \frac{(5)^2}{25} + \frac{(15)^2}{25} + \frac{(13)^2}{25} + \frac{(7)^2}{25}$$

$$= \frac{25}{25} + \frac{225}{25} + \frac{169}{25} + \frac{49}{25}$$

$$= 1.0 + 9.0 + 6.7 + 1.9$$

$$= 18.6$$

Step 6 Determine the degrees of freedom.

df = (Number of categories – 1)
df = (4 – 1)
df = 3

Step 7 Determine the significance of the calculated χ^2.

At df = 3, α of 0.05, $\chi^2 = 7.815$ for significance; the calculated χ^2 of $18.6 > 7.815$ and is significant at the 0.05 level; $p < 0.001$.

Step 8 Decide to reject or not reject the null hypothesis.

Because the calculated χ^2 is significant, the null hypothesis is rejected.

Step 9 Determine whether the statistical findings support the research hypothesis.

Because the null hypothesis was rejected, the research hypothesis that crickets would be differentially attracted to colors was supported.

Step 10 Construct a data table (see Table 11.9 *Attraction of Crickets to Various Colors*).

TABLE 11.9 Attraction of Crickets to Various Colors

Information	Observed distribution	Expected distribution (Chance)	Calculated χ^2
Mode	Blue	Red-Blue-Yellow-White	
Frequency distribution			
Red	30	25	1.0
Blue	40	25	9.0
Yellow	12	25	6.7
White	18	25	1.9
Number	100	100	
Results of the Chi-square test	$\chi^2 = 18.16$ at df = 3 χ^2 of $18.6 > 7.815$ $p < 0.001$		

Step 11 Write a paragraph describing the results.

The distribution of crickets on various colors, as compared with a chance distribution, is summarized in Table 11.9. More crickets appeared on red and blue and fewer on yellow and white than was predicted by chance. Chi-Square was used to test the following null hypothesis at the 0.05 level of significance:

The frequency distribution of crickets on various colors is not significantly different from the frequency distribution predicted by chance.

The null hypothesis was rejected ($\chi^2 = 18.6 > 7.815$ at df = 3; $p < 0.001$). The data supported the research hypothesis that crickets were differently attracted to the color blue and repelled by the color yellow.

Step 12 Write an appropriate conclusion.

The relative attraction of crickets to the colors red, yellow, blue and white was investigated by comparing the distribution of 100 crickets on the colors with a chance distribution. Significant differences existed, with crickets being repelled by the color yellow and attracted to the color blue. Although the research hypothesis was supported, the findings conflicted with the documented attraction of bees to yellow. Additional investigations should be conducted to determine if the frequency of the color used was a factor or if true interspecies differences occur. Adaptive benefits of the color preferences could also be determined including food gathering or prey avoidance.

Multiple chi-square comparisons must be used when more than one experimental group is involved in an experiment. As previously described for the *t* test, the probability of obtaining a significant value by chance increases with the number of comparisons. Therefore, chi-square tests should be reduced to the minimum needed to answer the research question, and the level of significance should be increased to 0.01 or above. Chi-Square tests should not be conducted when the observed frequency count is less than 5; categories may be combined to avoid these problems.

MULTIPLE COMPARISONS

Scenario: Steven read that the ratio of black to white moths in England increased over time as the environment became covered with soot. He designed a simulation for testing the effect of environmental changes on the type of offspring surviving. Newspaper was used as the background (environment) from which three types of paper prey (red, white, and newsprint) were randomly selected. Initially, 99 prey, 33 of each color, were placed upon the newsprint. Simulating a predator, Steven randomly selected 9 prey from the environment; remaining organisms were doubled to simulate reproduction and the selection process repeated. Prey of each color were counted and the relative number of each based upon 100 were expressed. The activity was repeated for the equivalent of four generations. Steven hypothesized that differential prey distribution would occur by the third generation. He used chi-square to determine if a significant difference in the distribution of prey occurred between Generations 0 and 1, Generations 0 and 2, Generations 0 and 3.

	Red prey	White prey	Newsprint prey
Generation 0	33.0	33.0	33.0
Generation 1	32.0	30.0	38.0
Generation 2	27.0	25.0	48.0
Generation 3	15.0	20.0	65.0

Step 1: Null hypotheses:

$1H_0$: Frequency Distribution of Moths, Generation 0 = Frequency Distribution of Moths, Generation 1

$2H_0$: Frequency Distribution of Moths, Generation 0 = Frequency Distribution of Moths, Generation 2

$3H_0$: Frequency Distribution of Moths, Generation 0 = Frequency Distribution of Moths, Generation 3

Step 2: Alpha (α) = 0.01.

Step 3: The observed frequency distributions for Generations 1,2, and 3 are stated in the scenario.

Step 4: The expected (theoretical) frequency distribution is equivalent to Generation 0, as stated in the scenario.

Step 5: Calculate χ^2.

$$\chi^2_{0-1} = \frac{(32-33)^2}{33} + \frac{(30-33)^2}{33} + \frac{(38-33)^2}{33}$$

$$= .03 + .27 + .76$$
$$= 1.05$$

$$\chi^2_{0-2} = \frac{(27-33)^2}{33} + \frac{(25-33)^2}{33} + \frac{(48-33)^2}{33}$$

$$= 1.09 + 1.94 + 6.82$$
$$= 9.85$$

$$\chi^2_{0-3} = \frac{(15-33)^2}{33} + \frac{(20-33)^2}{33} + \frac{(65-33)^2}{33}$$

$$= 9.82 + 5.12 + 31.0$$
$$= 45.9$$

Step 6: Degrees of Freedom = (3 – 1) = 2.

Step 7: Because at df = 2; α of 0.01, χ^2 = 9.210 for significance, the chi-square for the Generation 0 – 1 comparison is not significant. The chi-square for the Generation 0 – 2 and Generation 0 – 3 comparisons are significant.

Step 8: Do not reject

$1H_0$: Frequency Distribution of Moths, Generation 0 = Frequency Distribution of Moths, Generation 1

Reject

$2H_0$: Frequency Distribution of Moths, Generation 0 = Frequency Distribution of Moths, Generation 2

Reject

$3H_0$: Frequency Distribution of Moths, Generation 0 = Frequency Distribution of Moths, Generation 3

Step 9: The research hypothesis that differential prey distribution would occur by the third generation was supported. Significant differences occurred by the second generation.

Steps 10–12: Using the strategies previously described, design an appropriate data table and write appropriate results and a conclusion for Steven's study.

EVALUATING YOUR SKILLS

After you complete the appropriate statistical tests, *chi-square* or *t test,* rate your skills using Part One of Table 11.10 *Checklist for Evaluating Inferential Statistics.* Use Parts Two, Three, and Four to evaluate your data tables, paragraphs of results, and conclusion.

Related Web Site

http://www.mste.uiuc.edu/stat/stat.html

TABLE 11.10 Checklist for Evaluating Inferential Statistics

Criteria	Self	Peer/Family	Teacher
Part One—Inferential Statistical Test			
Correct null hypothesis			
Correct level of significance			
Correct statistical test			
Correct calculations			
Correct degrees of freedom			
Correct table value for statistic			
Correct interpretation of test—significance			
Correct action about null hypothesis			
Correct action about research hypothesis			
Part Two—Data tables			
Title			
Descriptive statistics			
Name of inferential statistical test			
Comparison of calculated/table values			
Degrees of freedom			
Significance/Probability level			
Part Three—Paragraphs; Results			
Topic Sentence			
Comparison of descriptive statistics			
Description of statistical test			
Interpretation of statistical test			
Support for research hypothesis			
Writing/Grammar/Spelling			
Part Four—Conclusion			
Purpose of experiment			
Major findings, including statistical test			
Support of research hypothesis by data			
Comparison with other research			
Explanation for findings			
Recommendations			
Writing/Grammar/Spelling			

INVESTIGATION 11.2 ■ Containing the Curdles

Question

How does temperature affect the curdling of milk?

Cold (ice)

Materials for Each Group of Two Students

150 ml of milk at room temperature
3 small clear zip bags—snack size
Ice—crushed
Beaker of hot water from faucet
3 bowls or other containers
1 graduated cylinder (100 ml)
1 graduated cylinder (10 ml) or
calibrated dropper
18 ml of white vinegar
Tape
Pen or pencil

Hot (faucet)

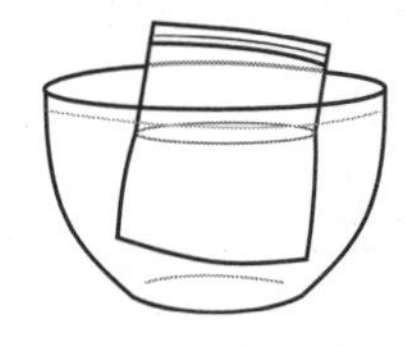

Room temperature (empty)

Safety

- Be careful handling the hot water. The temperature of water from faucets is very different.
- Wear goggles.
- Dispose of zip bags as directed by your teacher.
- Do NOT eat/drink the foods used in this experiment.
- Wash your hands after handling food/chemicals.

Procedure

1. Using tape, label each zip bag with one of the following labels: cold, room temperature, hot.
2. Add 50 ml of milk to each zip bag. Close it tightly.
3. Place the first bag of milk in a bowl and surround the bag with crushed ice. Place the second bag of milk in an empty bowl. Place the third bag of milk in a bowl and fill the bowl with hot water from the faucet. Leave the bags of milk in the bowl for 10 min.
4. Remove the bags from the bowls. Quickly unzip each bag and add 6 ml of vinegar to each bag. Reseal and return to the appropriate bowls. Describe the appearance of the curdles using the following symbols:

   SC—small curdles with milk generally not separated and still white in color
   MC—moderate curdles with milk beginning to separate and show yellow color
   LC—large curdles with milk separated and showing yellow color

   Note: To reduce odor, do not unzip the bags after you have added vinegar. Dispose of the bags as directed by your teacher.

(continued on the following page)

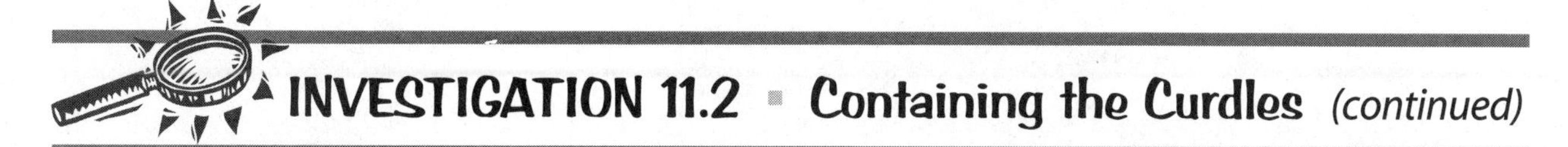

INVESTIGATION 11.2 ▪ Containing the Curdles *(continued)*

5. Enter the data in a class data table. Compile data from several classes, so that you have 50 or more trials.

Analyzing Your Data

1. Conduct a chi-square test to determine if significant differences exist among the appearance of the milk curdles at various temperatures, e.g. room vs. cold, room vs. hot, cold vs. hot.
2. Summarize your results with an appropriate data table, graph, and paragraph. (Hint: See Table 11.9 and accompanying paragraph.)
3. What is the chemical composition of milk? What causes the souring of milk? What factors affect the speed at which this chemical reaction occurs?
4. Do you think that you would get the same results with all types of milk? With various types of acids? How could you design experiments to test your hypotheses?

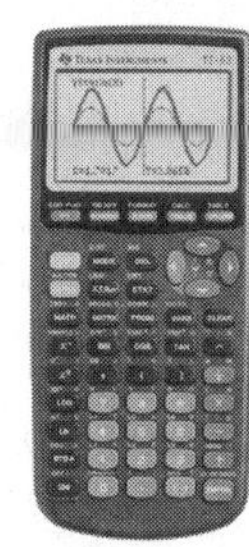

USING TECHNOLOGY

When you have* one *independent variable, as in this investigation:

Because graphing calculators only include a built-in program for a more complex chi-square test for two independent variables, you will need to enter a simple program for a one-independent variable chi-square test. Enter the program found in Appendix A, *Using Technology,* for a one-way chi-square calculation into your calculator. When prompted by the program, enter the number of categories, and for each category, enter the observed and expected values. When the calculated chi-square value is displayed, use Table 11.8 Chi-Square Sampling Distribution, to determine the probability that differences in frequencies were due to chance.

When you have* two *independent variables:

1. In the **MATRIX** mode of your calculator, create the size of the matrix needed for your data, for example, **2 x 3** for a study of the effect of gender (two, male and female) and preference for certain TV shows (three shows). (See Appendix A, *Using Technology*, for additional help in using the graphing calculator).
2. Enter the *observed* data in each cell of the matrix; the calculator will calculate the *expected* values. In some brands of calculators, you must create a second matrix for the *expected* values.
3. In the **STAT** mode of your calculator, select **TESTS** and choose chi-square.
4. Enter the letter name for the matrix or matrices, and then calculate.
5. The calculated chi-square value, the probability the differences in frequencies are due to chance, and the degrees of freedom will be displayed.

CHAPTER 12

Designing Complex Experiments

Objectives

- Diagram complex experimental designs including repeated measures over time or subjects, multiple independent variables, and correlation of variables.
- Describe positive features of complex experimental designs and select the most appropriate design for a given experiment.
- Construct appropriate data tables and graphs for data derived from complex experimental designs.
- Write appropriate paragraphs of results and conclusions for investigations involving complex experimental designs.

National Standards Connections

- Design and conduct scientific investigations (NSES).
- Design a statistical experiment to study a problem, conduct the experiment, and interpret and communicate the outcomes (NCTM).
- Use curve fitting to predict from data (NCTM).
- Transform data to aid in data interpretation and prediction (NCTM).

If you are a beginning researcher, experiments with one independent variable are recommended. Basic concepts of experimental design and ways to improve these designs are described in Chapters 1 and 2. With experience, you can handle more complex designs involving repeated measures over time or subjects, multiple variables, and the correlation of variables. For each of these designs, a scenario, an experimental design diagram, and graphical procedures for analyzing data are presented in this chapter.

REPEATED MEASURES OVER TIME

One simple way to modify and improve an experiment with one independent variable is to obtain multiple measures of the dependent variable over time. An experiment to determine the influence of earthworms on soil quality would be enhanced by reporting results weekly, rather than at the end of a two-month period. Similarly, the influence of aerobic exercise on resting pulse rate can be more accurately evaluated with monthly measurements, rather than one at the end of a year. This design is particularly effective when different effects over time are hypothesized. For example, one fertilizer may act more quickly than another to promote plant growth, yet they may produce equivalent growth at six weeks.

Scenario

Ronald raised chickens as a hobby. In his science class, he studied the importance of a balanced diet. The feed Ronald normally purchased contained a greater percentage of carbohydrates and fats and a lower percentage of protein than recommended. Ronald decided to mix his own feed to include the recommended daily allowances. He hypothesized that the chicks fed a balanced diet would have a smaller mass and be more active. Thirty-five chicks were fed the nonnutritionally balanced feed; an additional 35 chicks were fed the home-mixed nutritionally balanced feed. The chicks were the same age, received the same amount of food and water, and were reared under identical climate and space conditions. The mass and activity level of the chicks were measured weekly. Activity level was rated on a scale of 1 to 3, with 1 = inactive, 2 = moderately active, and 3 = very active. To ensure no harmful effects on the chicks, a local agricultural extension agent served as Ronald's mentor. (Note: All projects involving vertebrates must be approved in advance and supervised by a mentor; the experiment must be discontinued immediately if harmful effects occur.)

Design and Analysis

Use a modified experimental design diagram format, as illustrated in Table 12.1 Experimental Design Diagram, to communicate both the independent variable and the time intervals. Put the variable time across the top of the rectangle and the levels of the independent variable along the side. Although these can be switched, the data display will be easier if you show time horizontally.

You can change an experimental design diagram into a data table by adding a column for descriptive information that includes the appropriate measures of central tendency, variation, and number. Data tables for chick mass (quantitative) and activity rate (qualitative) are illustrated in Table 12.2 *Data Display: Diet and Chickens.* Because repeated measures are involved, the variables dis-

TABLE 12.1 Experimental Design Diagram

Title: The Effect of Diet on Chicken Growth and Behavior
Hypothesis: If chicks are fed a balanced diet, then they will have a smaller mass and be more active.

	Time (weeks)					
Type of diet	**1**	**2**	**3**	**4**	**5**	**6**
Non-balanced (Control)	35 chicks	35 chicks	35 chicks	35 chicks	35 chicks	35 chicks
Balanced (Experimental)	35 chicks	35 chicks	35 chicks	35 chicks	35 chicks	35 chicks

DV: Mass (g)
Activity (scale of 1 to 3)

C: Age
Food
Water
Climate
Space

TABLE 12.2 Data Display: Diet and Chickens

Diet & Mass
Data Table and Graph A—Effect of Diet on Chick Mass (g)

Type of diet	Descriptive information	Time (weeks) 1	2	3	4	5	6
Non-balanced diet	Mean	610	648	698	750	820	880
	Range	16	20	19	22	30	33
	Number	35	35	35	35	35	35
Balanced diet	Mean	612	630	660	680	720	745
	Range	18	19	21	18	21	19
	Number	35	35	35	35	35	35

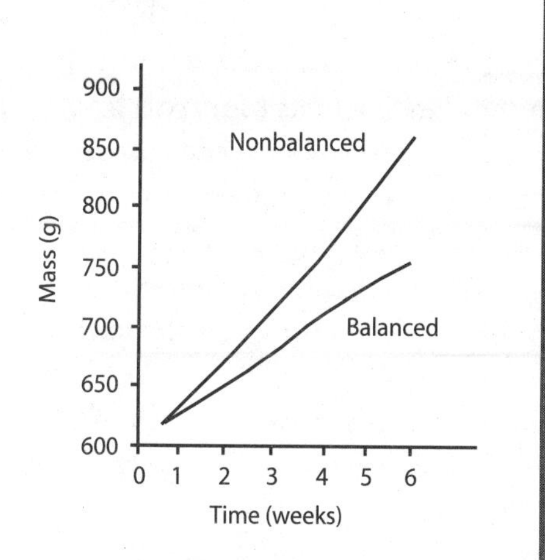

Diet & Activity
Data Table & Graph B—Effect of Diet on Chick Activity

Type of diet	Descriptive information	Time (weeks) 1	2	3	4	5	6
Nonbalanced diet	Median	3	3	2	2	1	1
	Frequency Distribution						
	Activity 3	20	25	10	8	5	6
	Activity 2	10	7	18	12	12	7
	Activity 1	5	3	7	15	18	22
	Number	35	35	35	35	35	35
Balanced diet	Median	3	3	2	2	2	2
	Frequency Distribution						
	Activity 3	20	23	17	14	11	10
	Activity 2	8	6	13	18	20	22
	Activity 1	4	6	5	3	4	3
	Number	35	35	35	35	35	35

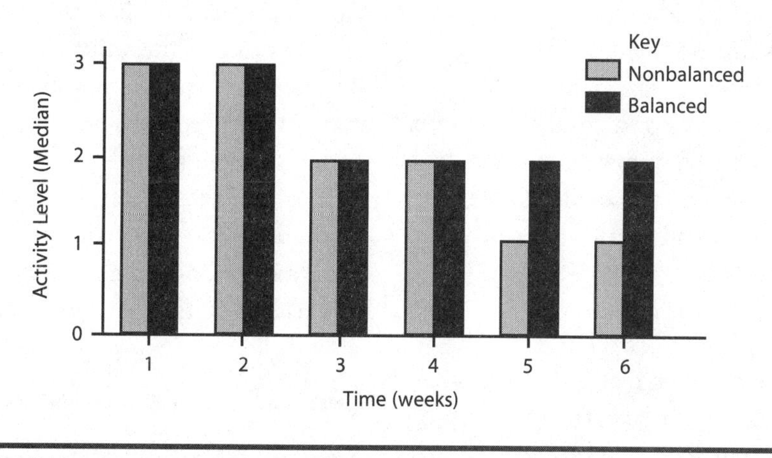

played on the axes of the graph must be changed. Display time intervals on the X axis and the dependent variable, mass, on the Y axis. Using different symbols, graph the mass of the chicks fed the nonbalanced feed (control) and then graph the mass of the chicks fed the balanced feed. Similarly, construct a bar graph to display the activity level of chicks fed the two diets. Use procedures described in Chapter 9, *Communicating Descriptive Statistics*, for writing results and conclusions. For statistical analysis, consult reference texts on the correlated *t* test or one-way analysis of variance with repeated measures.

Conducting and Analyzing an Experiment

Read Investigation 12.1, *Holding the Heat* (on page 179). Design a class experiment to test the effectiveness of three different insulators. Draw an experimental design diagram. Conduct the experiment and summarize the data using appropriate tables, graphs, and paragraphs.

REPEATED TREATMENTS OVER SUBJECTS

By exposing the same subjects to different treatments, you can minimize experimental error resulting from variations within subjects. Repeated treatment designs are particularly effective in psychological and biological studies involving higher organisms. For example, the most effective time for learning could be investigated by determining 60 subjects' rates of learning nonsense syllables at three different times during the day. With this design, each student serves as his or her own control. Thus, genetic and environmental factors are minimized. Because nonliving matter exhibits less variation, repeated treatment designs are less common in the physical and earth sciences.

Scenario

Although Sarah's classmates learned the advantages of studying in a quiet place, the majority of the students continued to complete homework while watching television or listening to music. Sarah hypothesized that the time needed to solve mathematics problems would increase with the number of stimuli in the environment. Sarah taped three 30-minute segments of television programs that were equivalent in conversation, music, and screen action. She developed four equivalent mathematics tests (20 items) on decimals and percents. Fifteen students were randomly selected from the class; parental permission to participate was secured. Students completed a mathematics test while exposed to no stimuli, to an auditory tape, to a video, and to a tape of the television programs. The form of the mathematics test, type of stimuli, and order of presentation of the stimuli were randomized. Time for completion was recorded. Tests were administered at the same time of day in identical test sites. The sound level of the auditory stimuli and the screen size of the visual stimuli remained constant.

Design and Analysis

As diagrammed in Table 12.3 *Experimental Design Diagram*, place the levels of the independent variable across the top of the experimental design diagram and the subjects along the side. List the dependent variables and constants below the rectangle.

Calculate the descriptive statistics for each treatment group, including measures of central tendency (mean) and variation (standard deviation or range). Show the results graphically. As illustrated in Table 12.4 *The Effect of Stimuli on Time (sec) to Solve Mathematical Problems*, individual subjects' measures are reflected in the raw data but not in the summary data table or graph. Repeated measures on subjects are an important control feature of the experimental design and are emphasized in the methods and materials section, rather than the results section of the research report. For statistical analysis, consult reference texts on the correlated *t* test or one-way analysis of variance with repeated measures.

Conducting and Analyzing an Experiment

Read Investigation 12.2, *Practice Makes Perfect* (on page 181). Design a class experiment to test

TABLE 12.3 Experimental Design Diagram

Title: The Effect of Stimuli on Mathematics Problem Solving
Hypothesis: Time for completion of mathematics tests will increase as the number of stimuli in the environment increases.

Subjects	Type of stimuli			
	None (Control)	Auditory	TV Tape	Video
1 2 3 • • • 15				

DV: Time to complete (sec)
C: Length and difficulty of test
Length, complexity, and intensity of stimuli
Time and place of administration

the impact of practice on the ability to perform a task. Draw an experimental design diagram. Conduct the experiment and summarize the data using appropriate tables, graphs, and paragraphs.

TWO INDEPENDENT VARIABLES

When the literature review suggests that two factors may interact to influence the outcome, an experimental design involving two independent variables is appropriate. For example, concentration and time of application may influence the effectiveness of a weed killer. In experiments with two independent variables, both main and interaction effects are hypothesized. **Main effects** refer to the action of each independent variable alone, whereas **interaction** refers to the combined action of the variables. Experiments with two independent variables are not appropriate unless an interaction is hypothesized.

- **Main Effect:** Higher concentrations of Herbicide X will result in a greater death rate of weeds.
- **Main Effect:** Early morning application of Herbicide X will result in a greater death rate of weeds.
- **Interaction:** Time of application and concentration will interact to affect the death rate of weeds, for example, lower concentrations of Herbicide X will effectively kill weeds in the morning while higher concentrations will be required in the afternoon.

Scenario

Larry read several articles about the effects of playing video games. One article indicated that playing video games increased an individual's stress level. A second article suggested that older people displayed more stress than younger people. The third article indicated that stress level was a function of the type of video game and the relevance of the game to real-life situations. Larry decided to investigate the effect of both age and type of video game on stress level as measured by the pulse rate of the individual. Larry randomly selected 45 subjects (SS) subdivided into three age groups: (a) 10–12 years (15 SS), (b) 15–17 years (15 SS), and (c) 20–23 years (15 SS). Five subjects

TABLE 12.4 The Effect of Stimuli on Time (sec) to Solve Mathematics Problems

Data table for raw data:

Subjects	Type of stimuli			
	None	**Auditory**	**TV Tape**	**Video**
1	20.0	35.0	21.0	36.0
2	17.0	20.0	16.0	22.0
3	22.0	27.0	24.0	30.0
4	14.0	18.0	15.0	18.0
5	13.0	19.0	14.0	18.0
6	25.0	30.0	25.0	29.0
7	18.0	28.0	22.0	30.0
8	12.0	15.0	11.0	14.0
9	15.0	20.0	19.0	20.0
10	16.0	21.0	19.0	24.0
11	21.0	30.0	23.0	29.0
12	17.0	23.0	19.0	24.0
13	18.0	20.0	19.0	21.0
14	20.0	28.0	22.0	29.0
15	15.0	24.0	19.0	28.0

Data table for summarized data:

Descriptive information	Type of stimuli			
	None	**Auditory**	**TV Tape**	**Video**
Mean	17.5	23.9	19.2	24.8
Range	13.0	20.0	14.0	22.0
Maximum	25.0	35.0	25.0	36.0
Minimum	12.0	15.0	11.0	14.0
Number	15	15	15	15

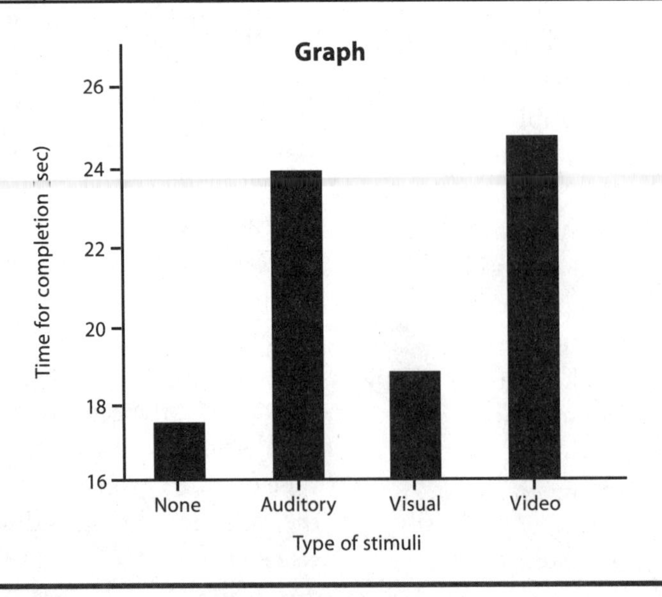

within each age group were then assigned to play a video game: (a) Wac-Man (5 SS), (b) Busdec (5 SS), and (C) Cosmos (5 SS). Each individual played the game in the same setting for 20 minutes. Larry hypothesized that individuals would place greater importance on winning Wac-Man, a popular recreation game. Successively less importance would be placed upon winning Busdec, a business simulation, and Cosmos, a space game. He also hypothesized that age and type of video game would interact with older students exhibiting greater stress with Busdec. Larry stated three research hypotheses to guide the experimental design and analysis:

- **Main Effect:** The age of the subject will influence pulse rate after playing a video game with the oldest subjects exhibiting the highest pulse rate.
- **Main Effect:** The type of video game will influence the pulse rate of subjects, such as Wac-Man > Busdec > Cosmos.
- **Interaction:** The subject's age and type of video game will interact to influence the subject's pulse rate, for example, younger subjects will exhibit the highest pulse rate with Wac-Man, and older subjects with Busdec. Reaction to Cosmos will be equivalent across age groups.

Design and Analysis

Modify the experimental design diagram to show one independent variable across the top and the second along the side (see Table 12.5 *Experimental Design Program*). Generally, the treatment variable is shown across the top and a categorical variable along the side, such as age, intelligence, gender. Remember to include the dependent variables, constants, and number of repeated trials in the experimental design.

To determine the main and interactive effects of the two variables, measures of central tendency and variation must be computed for each cell, each horizontal row, and each vertical column of the raw data table. Construct a reduced data table from the raw data table as described next and illustrated in Table 12.6 *The Effect of Age & Video Game on Pulse Rate (Beats/Minute)*. For statistical analysis, consult a reference text on two-way analysis of variance.

TABLE 12.5 Experimental Design Diagram

Title: The Effect of Age and Type of Video Game on Pulse Rate
Hypothesis:

- Older subjects will exhibit higher pulse rates.
- Type of video game will affect pulse rate.
- Subject's age and type of video game will interact to affect pulse rate.

	IV: Type of Video Game		
IV: Age group	**Wac-Man**	**Cosmos**	**Busdec**
10–12 years	5 people	5 people	5 people
15–17 years	5 people	5 people	5 people
20–23 years	5 people	5 people	5 people

DV: Change in pulse rate
C: Time of play (20 min)
Setting

TABLE 12.6 The Effect of Age and Video Game on Pulse Rate

Data table for raw data:

	Type of video game		
Age group	**Wac-Man**	**Busdec**	**Cosmos**
10–12 years	80.0 90.0 82.0 85.0 78.0	82.0 88.0 84.0 87.0 80.0	90.0 98.0 100.0 88.0 85.0
15–17 years	78.0 94.0 84.0 81.0 80.0	88.0 95.0 80.0 82.0 85.0	80.0 88.0 76.0 83.0 80.0
20–23 years	88.0 90.0 87.0 95.0 94.0	78.0 75.0 70.0 72.0 73.0	60.0 62.0 64.0 68.0 62.0

Data table for summarized data:

	Type of video game			
Age group	**Wac-Man**	**Busdec**	**Cosmos**	
10–12 years	$\bar{X}$ = 83.0 SD = 4.2 n = 5	$\bar{X}$ = 84.2 SD = 2.9 n = 5	$\bar{X}$ = 92.2 SD = 5.8 n = 5	$\bar{X}$ = 86.5 SD = 6.1 n = 15
15–17 years	$\bar{X}$ = 83.4 SD = 5.6 n = 5	$\bar{X}$ = 86.0 SD = 5.3 n = 5	$\bar{X}$ = 81.4 SD = 4.0 n = 5	$\bar{X}$ = 83.6 SD = 5.3 n = 15
20–23 years	$\bar{X}$ = 90.0 SD = 3.2 n = 5	$\bar{X}$ = 73.6 SD = 2.7 n = 5	$\bar{X}$ = 63.2 SD = 2.7 n = 5	$\bar{X}$ = 75.9 SD = 11.7 n = 15
	$\bar{X}$ = 85.7 SD = 5.7 n = 15	$\bar{X}$ = 81.3 SD = 6.7 n = 15	$\bar{X}$ = 78.9 SD = 13.3 n = 15	$\bar{X}$ = 81.9

- **Main Effect for Type of Video Game:** Treat the data as if they were from an experiment with one independent variable (type of video game) by calculating the mean and standard deviation for each vertical column. Place the summary statistics under the appropriate column.
- **Main Effect for Influence of Age:** Treat the data as if they were from an experiment with one independent variable (age) by calculating the mean and standard deviation for each horizontal row. Record the summary statistics for each row in the right-hand margin of the table.
- **Interactive Effect of Type of Video Game and Age:** Compute the mean and standard deviation for each cell within the experimental design.

Tables and Graphs

Methods for constructing data tables and graphs to display main effects are described in Chapters 5 and 8. The interaction of the variables is also determined by graphical analysis. One independent variable (for example, type of video game) is depicted along the X axis and the dependent variable on the Y axis. The mean pulse rate for the individuals, aged 10–12 years, is graphed. Using different symbols, data for the other two age groups are displayed. No interaction exists if the patterns of lines on the graph are the same. An interaction exists when graphed lines cross or diverge from parallel alignment. Similarly, an interaction exists when sets of bars on a graph have different patterns. A bar graph is the appropriate graph for Larry's data because type of video game is categorical data and the intervals between the levels of the independent variable are meaningless (see Chapter 5). In Larry's experiment, an interaction occurred between age and type of video game. Pulse rate of ages 10–12 was less when playing Wac-Man, while pulse rates of ages 20–23 were greater. The pulse rate pattern for ages 20–23 is opposite the observed pulse rates for the other two age groups. (See Table 12.7 *Data Display, Age and Video Games on Pulse Rate* for illustrative data tables and graphs.) When writing the results, the findings for each of the main effects and the interaction must be described.

Results

The effect of type of video game on pulse rate is summarized in Table and Graph 12.7A. Pulse rate was highest in Wac-Man players (85.7), followed by Busdec (81.3), and Cosmos (78.9) players. Variance in pulse rate was greater among Cosmos players. The hypothesized influence of type of video game on pulse rate was supported.

The influence of age on pulse rate is summarized in Table and Graph 12.7B. Substantially lower pulse rates occurred in ages 20–23 (75.9) than among individuals aged 15–17 years (83.6) or 10–12 years (86.5). Variation in pulse rates was greatest among ages 20–23. The data contradicted the research hypothesis that older subjects would exhibit progressively higher pulse rates.

Interactive effects of the subject's age and type of video game are summarized in Table and Graph 12.7C. Higher pulse rates occurred among older subjects (20–23 years) playing Wac-Man and younger subjects (10–12; 15–17 years) playing Cosmos. The experimental data do not support the hypothesized pulse increase with older subjects playing Busdec or younger subjects playing Wac-Man. Substantial differences were found among the pulse rates of different aged subjects playing Cosmos. The data do not support the hypothesis that Cosmos players would have equivalent pulse rates. Instead, substantial interaction occurred with subjects aged 20–23 exhibiting the highest pulse rate with Wac-man, subjects aged 15–17 the highest pulse rate with Busdec, and subjects aged 10–12 the highest pulse rate with Cosmos. In general, these findings support the hypothesis of an interactive effect of the subject's age and type of video game on pulse rate, but not the specifics of the interaction.

THREE INDEPENDENT VARIABLES

Experiments involving three independent variables are appropriate when three factors may potentially interact to influence the value of the de-

TABLE 12.7 Data Display, Age and Video Games on Pulse Rate

A
Effect of Type of Video Game on Pulse Rate (beats/minute)

Descriptive information	Type of video game		
	Wac-Man	**Busdec**	**Cosmos**
Mean	85.7	81.3	78.9
Standard deviation	5.7	6.7	13.3
Number	15	15	15

B
Effect of Age on Pulse Rate (beats/minute)

Descriptive information	Age Group		
	10–12 yrs.	**15–17 yrs.**	**20–23 yrs.**
Mean	86.5	83.6	75.9
Standard deviation	6.1	5.3	11.7
Number	15	15	15

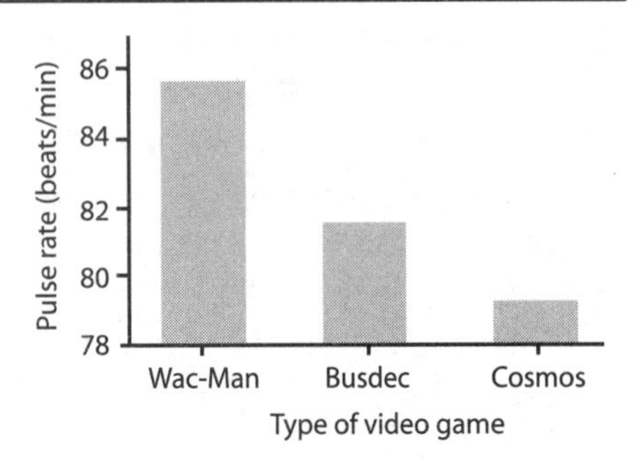

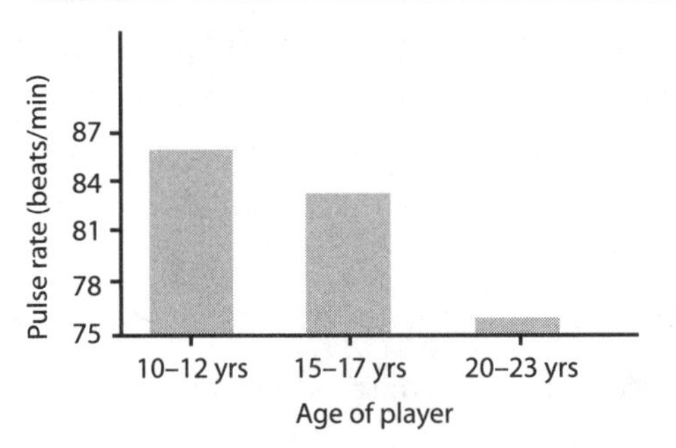

C
Effect of Age and Type of Video Game on Pulse Rate (beats/minute)

Age group	Type of video game		
	Wac-Man	**Busdec**	**Cosmos**
10–12 years	$\bar{X}$ = 83.0 SD = 4.2	$\bar{X}$ = 84.2 SD = 2.9	$\bar{X}$ = 92.2 SD = 5.8
15–17 years	$\bar{X}$ = 83.4 SD = 5.6	$\bar{X}$ = 86.0 SD = 5.3	$\bar{X}$ = 81.4 SD = 4.0
20–23 years	$\bar{X}$ = 90.8 SD = 3.2	$\bar{X}$ = 73.6 SD = 2.7	$\bar{X}$ = 63.2 SD = 2.7

n = 5 subjects/cell

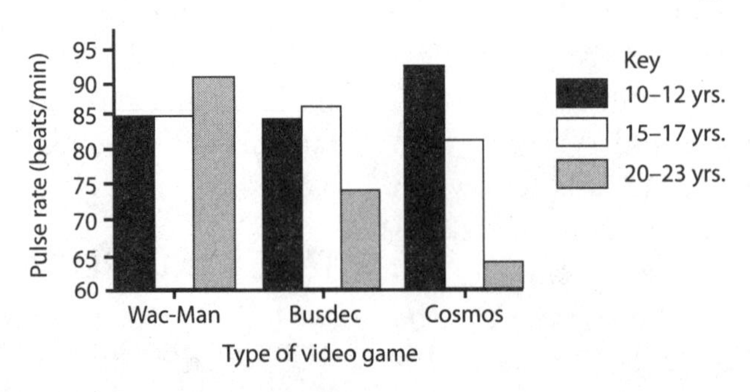

pendent variable. Respiratory rate in fish, for example, may be affected more strongly by the combination of thermal, phosphate, and acid rain pollution than would be predicted by the action of each variable alone. When an experiment includes three independent variables, then three main effects, three double interactions, and one triple interaction can be investigated. Experimental designs with three independent variables are appropriate only when a triple interaction is hypothesized.

Main Effects	Thermal Pollution Phosphate Pollution Acid Rain Pollution
Double Interactions	Thermal and Phosphate Pollution Thermal and Acid Rain Pollution Phosphate and Acid Rain Pollution
Triple Interaction	Thermal, Phosphate, and Acid Rain Pollution

Scenario

Nina was interested in the effectiveness of aerosol and nonaerosol forms of Superwash and Out! prewash cleaners in removing common stains, such as catsup, mustard, grass, new motor oil, soot, and used motor oil. She hypothesized that:

- **Main Effect:** Plant stains (catsup, mustard) will be easier to remove than hydrocarbon stains (new motor oil, soot, used motor oil).
- **Main Effect:** Superwash will remove stains more effectively than Out! or the detergent alone (control).
- **Main Effect:** Aerosol prewashes will remove stains more effectively than nonaerosol prewashes.
- **Interaction:** Nonaerosol types of Superwash will be more effective on plant stains, aerosol forms of Out! will be more effective on hydrocarbon stains.

Cotton-polyester sheeting was cut into 15 cm X 15 cm squares and a ring drawn in the center. One scoop of each stain was applied, evenly spread over the circle, and allowed to dry for 24 hours. Nine samples of each stain were treated with aerosol Out!, nonaerosol Out!, aerosol Superwash, and nonaerosol Superwash. Nine samples of each stain remained untreated and served as a control. All samples were washed on the regular cycle with warm water. The recommended amount of Brand X detergent was used. After drying for 24 hours, the degree of stain removal was rated on a 5-point scale, with a score of 5 representing total removal and 1 representing no removal.

Design and Analysis

In Table 12.8 *Experimental Design and Raw Data*, a three-way experimental design is illustrated; stain removal ratings for the nine samples/cell are included so data analysis will be easier for each section. You should show one independent variable (type of stain) across the top of the rectangle and the remaining two variables (type and brand of prewash) along the side. The number of trials is noted within each cell.

In the experiment, Nina used an ordered categorical scale (1–5) to rate cleaning effectiveness. Thus, the appropriate measure of central tendency is the median; variation in data is expressed through a frequency distribution. Begin summarizing data by computing a frequency distribution and median for each cell. Then, compute the frequency distribution and median value for:

- All samples washed within a stain category, e.g., mustard, catsup . . . used oil *(Place the value under the appropriate vertical column);*
- All samples washed with aerosol and all samples washed with nonaerosol prewash *(Place the value in the far right margin of the table);*

TABLE 12.8 Experimental Design Diagram and Raw Data

The Effect of Types of Stains, Prewash Brands, and Prewash Types on Stain Removal (Scale of 1 to 5)

- Plant stains (catsup, mustard) will be easier to remove than hydrocarbon stains (new motor oil, soot, used motor oil).
- **Superwash** will remove stains more effectively than **Out!** or the detergent alone (control).
- Nonaerosol types of **Superwash** will be more effective on plant stains, while aerosol forms of **Out!** will be more effective on hydrocarbon stains.

IV: Type of stain (mustard, catsup, grass, new oil, soot, used oil)
Brand of prewash (**Superwash, Out!**, none)
Type of prewash (aerosol, nonaerosol, none)

Prewashes		Stains					
Types	**Brands**	**Mustard**	**Catsup**	**Grass**	**New Motor Oil**	**Soot**	**Used Motor Oil**
Aerosol	Superwash	4 4 4 5 3 5 5 5 4	4 5 4 5 5 5 4 5 4	4 3 5 5 4 4 3 4 4	2 3 2 3 3 3 2 3 3	1 2 2 2 2 3 3 3 3	2 1 3 2 2 2 2 3 2
	Out!	5 5 4 4 4 4 4 4 5	3 5 4 4 4 3 3 4 5	3 3 3 2 4 2 3 2 3	4 3 2 4 4 3 3 4 4	3 3 3 2 4 2 3 3 3	3 4 3 4 3 4 4 4 4
Nonaerosol	Superwash	5 4 4 5 3 5 5 5 5	5 5 4 4 5 5 4 5 5	4 4 4 3 2 3 3 3 3	2 3 2 3 2 2 2 3 3	2 1 3 2 3 2 2 2 2	2 3 2 2 2 3 3 2 1
	Out!	4 5 5 5 4 4 5 4 4	5 3 4 4 4 3 4 5 4	2 3 3 2 1 2 2 2 2	2 3 4 2 3 2 3 3 2	3 2 4 4 3 3 3 3 3	3 3 2 2 3 3 2 2 3
None	None (control)	4 2 4 4 3 3 3 3 3	3 4 3 3 4 3 4 3 4	2 3 3 2 3 3 3 2 3	2 1 2 1 1 1 1 2 1	1 3 1 1 2 2 2 1 1	2 1 1 3 1 1 1 2 2

DV: Amount of stain removed (Scale of 1 to 5, 5 = total removal and 1 = no removal)

C: Duration of wash
Temperature of water
Wash cycle
Amount of stain
Fabric stained
Drying time

- All samples washed with Superwash and all samples washed with Out! *(Place the value in the far right margin of the table);*
- All samples with no prewash—the control; *(Place the value in the right margin of the table).*

A reduced data table with median values for each cell and for each independent variable is provided in Table 12.9 *Data Analysis, Prewashes and Srain Removal (Scale of 1–5)*. For statistical analysis, consult a reference text on three-way analysis of variance.

Tables and Graphs

For each of the independent variables, construct a data table and graph. From the data displays (see Table 12.10 *The Effect of Types of Stains, Prewash Brands, and Prewash Types on Stain Removal (Scale of 1 to 5)*), determine the effect of the three variables acting alone.

- Effect of Type of Stain on Cleaning Effectiveness
- Effect of Brand of the Prewash on Cleaning Effectiveness
- Effect of Type of Prewash on Cleaning Effectiveness

To determine the interaction of variables, show one independent variable (for example, type of stain) on the X axis and the dependent variable (such as cleaning effectiveness) on the Y axis. Using two different sets of symbols, graph the cleaning effectiveness of aerosol Superwash and Out! Repeat the process for the nonaerosol sprays and the control (see Table 12.10). Describe the effect of each variable acting alone and in combination. Relate the findings to the research hypotheses.

Results

The influence of type of stain on cleaning effectiveness is illustrated in Table and Graph 12.10A. The highest median cleaning (4) occurred with mustard and catsup stains, followed by grass/new oil (median = 3), and soot/used oil (median = 2). In general, the research hypothesis that plant stains would be more effectively cleaned than hydrocarbon stains was supported. The exception was grass, which equaled new oil in cleaning difficulty. As depicted in Table and Graph 12.10B,

TABLE 12.9 Data Analysis, Prewashes and Stain Removal (Scale of 1 to 5)

Prewashes		Stains						
Types	Brands	Mustard	Catsup	Grass	New Motor Oil	Soot	Used Motor Oil	
Aerosol	Superwash	4	5	4	3	2	2	M = 3
	Out!	4	4	3	4	3	4	
Non-aerosol	Superwash	5	5	3	2	2	2	M = 3
	Out!	4	4	2	3	3	3	
None (Control)	None (Control)	3	3	3	1	1	1	$M_s = 3$ $M_o = 3$ M = 2
		M = 4	M = 4	M = 3	M = 3	M = 2	M = 2	

TABLE 12.10 The Effect of Types of Stains, Prewash Brands, and Prewash Types on Stain Removal (Scale of 1 to 5)

A
Main Effect, Type of Stain

Descriptive information	Type of stain					
	Mustard	Catsup	Grass	New oil	Soot	Used oil
Median	4	4	3	3	2	2
Frequency distribution						
5	17	15	2	0	0	0
4	20	20	9	6	3	6
3	7	10	20	17	19	14
2	1	0	13	16	16	18
1	0	0	1	6	7	7
Number	45	45	45	45	45	45

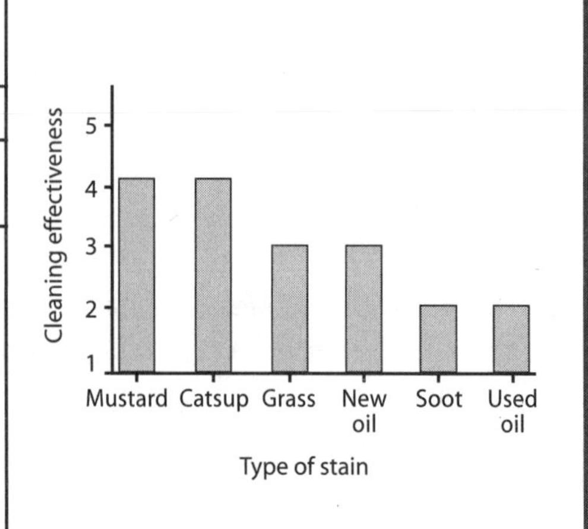

B
Main Effect, Brand of Prewash

Descriptive information	Brand of prewash		
	Superwash	Out!	None (Control)
Median	3	3	2
Frequency distribution			
5	23	11	0
4	21	36	7
3	30	39	17
2	30	21	14
1	4	1	16
Number	108	108	54

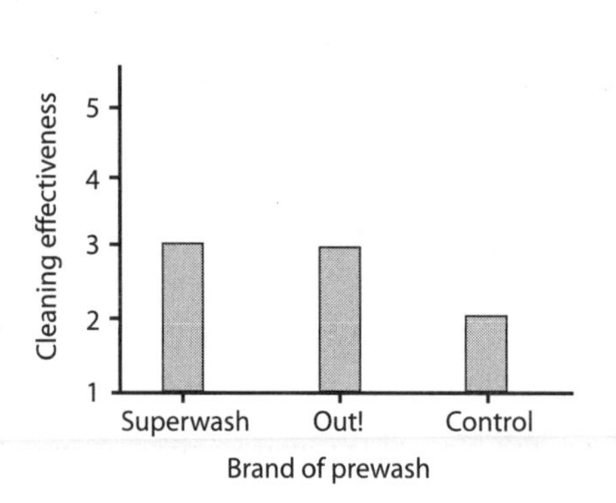

C
Main Effect, Type of Prewash

Descriptive information	Type of prewash		
	Aerosol	Nonaerosol	None (Control)
Median	3	3	2
Frequency distribution			
5	16	18	0
4	36	21	7
3	35	35	18
2	19	31	12
1	2	3	16
Number	108	108	54

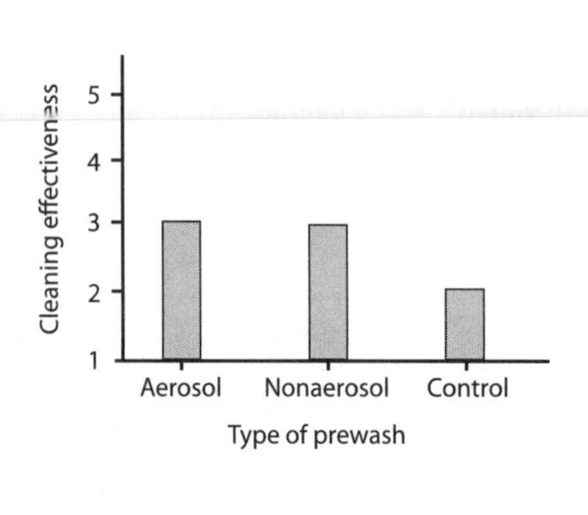

(continued on the following page)

TABLE 12.10 (continued)

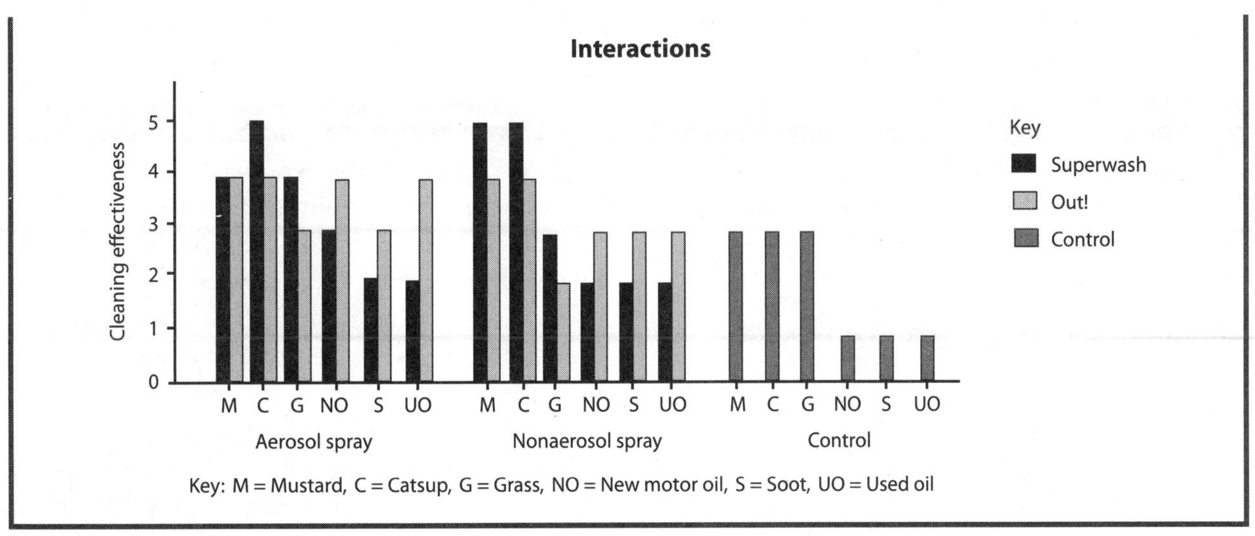

both brands of prewash were more effective than the control. Equivalent cleaning occurred with Superwash and Out! The hypothesized superiority of Superwash was not supported by the data. Aerosol and nonaerosol cleaners were equally effective in removing stains (see Table 12.10C). The data did not support the hypothesized superiority of aerosol prewashes. The type and brand of prewash influenced cleaning effectiveness. Aerosol and non-aerosol Superwash were more effective with plant stains and aerosol and nonaerosol Out! with hydrocarbons. The following lists indicate the most effective cleaner for each type of stain:

Stain	Most Effective Cleaners
Mustard	Nonaerosol Superwash
Catsup	Nonaerosol or aerosol Superwash
Grass	Nonaerosol Superwash
New oil	Aerosol Out!
Soot	Aerosol or nonaerosol Out!
Used Oil	Aerosol Out!

When both the type and brand of cleaner were considered, nonaerosol Superwash proved superior with plant stains and aerosol Out! with hydrocarbon stains. These findings supported the research hypothesis of differences in effectiveness of type and brand of prewash with various stains.

Conclusion

In general, plant stains proved easier to clean than hydrocarbon stains. Based upon chemical structure, these findings were predicted. Plants consist of complex carbohydrates containing oxygen, carbon, and hydrogen atoms; charged areas occur along the molecule. Hydrocarbons consist of long chains of hydrogen and carbon, which tend to repel both water and detergents; the molecule is nonpolar and does not have charged sections. The charged areas in carbohydrates promote bonding among the stain, detergent, and water, and thereby make stain removal easier. Prewash treatments proved more effective than the detergent alone (control). Both types and brands were equivalent in general stain removal.

Nonaerosol Superwash proved most effective with plant stains, and aerosol Out! with hydrocarbon stains. Prewashes act by promoting the breakdown of the stain and increasing the bonding of the detergent and water with the stain. Because substances with similar bonding are more soluble, Superwash must contain more charged (polar) particles while Out! contains more noncharged particles. The foaming agent may also increase the solubility of hydrocarbon stains, or greater wetting with nonaerosol sprays may promote removal of plant stains. The experiment could be improved by including more brands and by increasing the sample size. Additional studies should focus on stains that did not readily fit the pattern, for example, grass and new motor oil.

Conducting and Analyzing an Experiment

Read Investigation 12.3, *Fruit Appeal* (see page 183). Design a class experiment to determine the impact of various factors on the oxidation of fruit. Draw an experimental design diagram. Conduct the experiment and summarize the data using appropriate tables, graphs, and paragraphs.

CORRELATION

Researchers are frequently concerned with predicting the value of one variable (Y) from knowledge of a second variable (X). For example, college admissions officers predict success in college from SAT scores, and astronomers predict disturbances in radio transmission from the number of sunspots. Today, climatologists seek to predict the warming of the Earth's atmosphere from hydrocarbon levels. Prediction of one variable from another variable is possible only when a correlation exists.

If two variables, X and Y, are not related, a zero correlation exists. Knowledge of X tells the researcher nothing about Y, and vice versa. When the values of the two variables are graphed, as illustrated in Figure 12.1, no systematic pattern results. Systematic patterns emerge when a nonzero correlation exists. If the value of one variable (Y) increases with the value of the second variable (X), a positive correlation exists. The more closely data points approach a linear pattern, the stronger the correlation. When the value of one variable decreases (Y) as the value of the second variable increases (X), a negative correlation exists. The stronger the correlation between the variables, the more accurately one variable can be predicted from knowledge of the other (see Figure 12.1). Mathematical techniques, as well as graphical, also exist for expressing the correlation between variables. Computed values range from –1 to +1. If the correlation between variables is either +1 or –1, perfect prediction is possible. For statistical analysis of data, consult a reference text for Pearson's Product Moment Correlation or Spearman Rank Order Correlation.

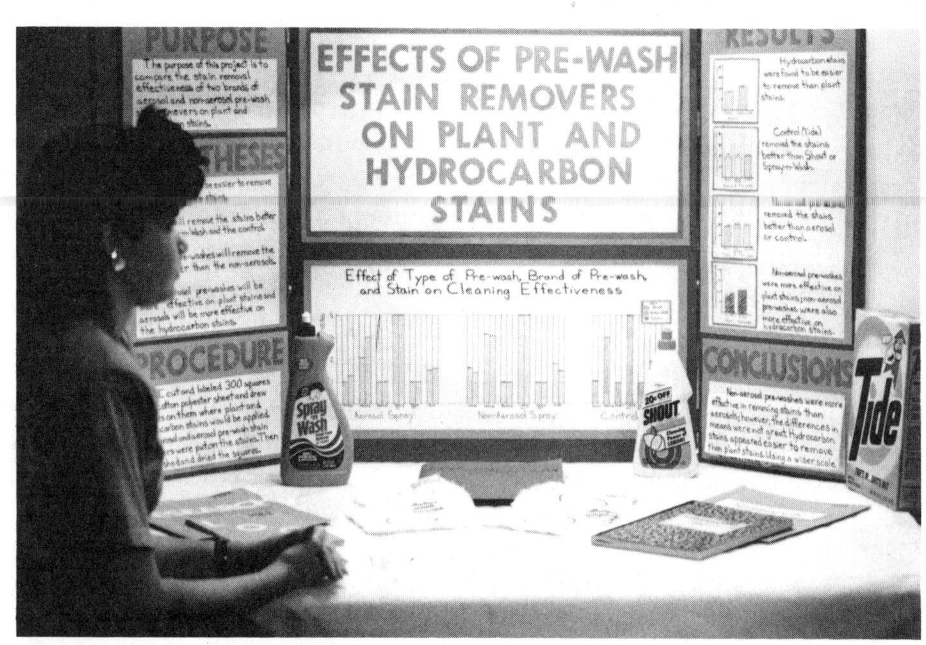

Photograph by Bruce Maxwell, print by Eleanor and Wilton Tenney.

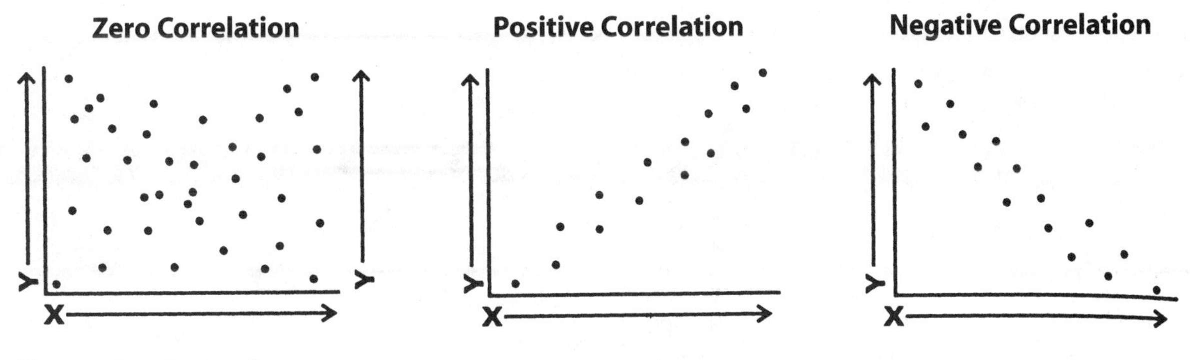

Figure 12.1 Scatterplot.

Scenario

Rita Lynn Gilman determined the wood production of loblolly pine trees over a 40-year interval. A major interstate was constructed next to the forest during the third and fourth decades. Rita Lynn had previously determined that wood production in that forest was significantly less than in a comparable forest fronted by rural roads. Because wood production substantially declined after construction of the interstate, Rita Lynn wondered whether wood production could be predicted from traffic counts.

Design and Analysis

In correlation studies, the value of each variable (X and Y) is simultaneously determined for each subject, year, and so on. For example, saturated fat consumption and blood cholesterol level could be determined for each of 5,000 subjects or the number of sunspots and intensity of northern lights could be secured from 50 years of records. The experimental design diagram shows the framework for the experiment (year, subject, and so on) and the specific variables measured. One variable (X) is designated as the independent (predictor) variable and the second as the dependent variable (Y). In Rita Lynn's experiment, the experimental framework was the 18-year interval. For each year, traffic count acted as the independent (predictor) variable and wood production as the dependent variable. Table 12.11 *Raw Data on Traffic Count and Tree Ring Width* contains the raw data.

TABLE 12.11 Raw Data on Traffic Count and Tree Ring Width

	Independent Variable	Dependent Variable
Year	Traffic Count (Vehicles/year)	Wood Production Tree Ring Area (mm²)
1965	26,240	1149.42
1966	26,010	1011.50
1967	27,325	966.680
1968	30,010	819.769
1969	30,090	931.373
1970	28,795	824.337
1971	30,740	864.514
1972	32,930	1021.28
1973	33,940	902.085
1974	31,690	866.515
1975	34,040	932.353
1976	35,540	869.129
1977	38,440	649.712
1978	41,760	605.717
1979	43,150	506.946
1980	41,070	560.024
1981	42,330	617.351
1982	43,700	437.180
1983	44,140	348.047

Determine an appropriate scale for depicting the independent variable (for example, traffic count) on the X axis and the dependent variable (for example, wood production) on the Y axis. Plot each data point on the graph. Examine the scatterplot to determine whether a zero, positive, or negative correlation exists. In this instance, a negative correlation existed (see Figure 12.2), because wood production decreased as traffic count increased.

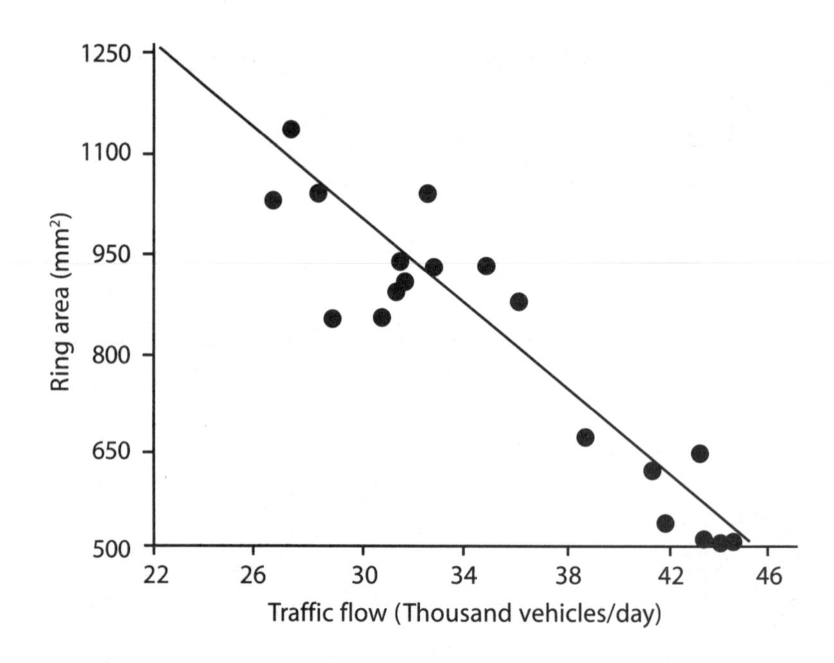

Figure 12.2 Relationship of Traffic Flow to Yearly Wood Production.

Conducting and Analyzing an Experiment

Read Investigation 12.4, *Collapsing Bridges* (see page 184). Design a class experiment to investigate the impact of bridge thickness on collapsing point. Draw an experimental design diagram. Conduct the experiment and summarize the data using appropriate tables, graphs, and paragraphs.

REFERENCES

Gilman, R. L. *The possible effects of pollutants on the wood production of Loblolly pine trees.* Paper presented at the meeting of the Virginia Junior Academy of Science, First Place, Environmental Science A & American Academy of Science Presentor, Harrisonburg, VA: James Madison University.

Glenberg, A. (1996). *Learning from data: An introduction to statistical reasoning* (2nd ed.). Mahwah, NJ: Lawrence Erlbaum Associates.

Landwehr, J.M. & Watkins, A.E. (1994). *Exploring data.* A component of the Quantitative Literacy Series. Palo Alto, CA: Dale Seymour Publications.

Landwehr, J.M., Swift, J., & Watkins, A.E. (1998). *Exploring surveys and information from samples.* A component of the Data-Driven Mathematics Series. Palo Alto, CA: Dale Seymour Publications.

Lyon, C. G. *The synergistic effects of acid rain, thermal pollution, and phosphate pollution on common shiner minnows (Notropis cornutus).* Paper presented at the meeting of the Virginia Junior Academy of Science, First Place, Environmental Science B and R. J. Rowlett Award for Best Research Paper, Harrisonburg, VA: James Madison University.

McClave, J.T., Dietrich, Franck H. II, Sincich, T. (1997). *Statistics* (7th ed.). Upper Saddle River, NJ: Prentice-Hall, Inc.

Neurohr, J.L. *The effect of solutions of various pH levels on the growth of kudzu (Pueraria thunbergiana).* Paper prepared for chemistry class at Lee Davis High School, Mechanicsville, VA.

Raybourne, K.D. *Heart girth as an indicator of live weight in white-tailed deer (Odocoileus virginianus).* Paper presented at the Virginia Junior Academy of Science, Second Place, Zoology, Harrisonburg, VA: James Madison University.

Shavelson, R.J. (1996). *Statistical reasoning for the behavioral sciences* (3rd ed.). Boston: Allyn & Bacon.

Yates, D.S., Moore, O.S., McCabe, G.P. (1999). *The Practice of Statistics: TI-83 Graphing Calculator Enhanced.* New York: W.H. Freeman and Company.

Related Web Sites

http://www.dade.k12.fl.us/us1/science/prod03.htm

http://www.mcrel.org/resources/links/index.asp

http://www.eduzone.com/Tips/science/SHOWTIP2.HTM (report section)

http://webster.commnet.edu/mla.htm

http://webster.commnet.edu/apa/apa_index.htm

http://155.43.225.30/workbook.htm

INVESTIGATION 12.1 ▪ Holding the Heat

Question

Which substance makes the best insulator?

Materials for Each Group of 4 Students

3 Aluminum soft drink cans with pull-off tabs
3 Larger cans such as used for vegetables or fruit
3 Thermometers (°C)
900 ml of hot water (80°C)
Insulated container to hold 1000 ml water
Device to heat water such as hot plate or Bunsen burner
Timer
Various insulators such as fiberglass, foam plastic packing foam, sawdust, cotton, air, shredded paper, other commercial products

Safety

- Wear goggles.
- Handle hot materials carefully.

Procedure

Using the general procedures described below, design a class experiment to test the effectiveness of 3 different insulators over a 30-minute period, collecting measurements at regular intervals. Draw the experimental design diagram for the experiment. Write detailed procedures that each lab group can follow. Design a class data table for compiling group data.

- Heat 1 liter of water to a temperature below the boiling point (approximately 80°C).
- Place 3 cm of insulator in the bottom of the larger vegetable can.
- Set the aluminum soft drink can on top of the insulation, being careful to center it.
- Pack the space between the larger vegetable can and the aluminum soft drink can with the insulator.
- Pour a specific amount of water (about 300 ml) into the aluminum soft drink can.
- Insert the thermometer into the water and begin recording the water temperature at regular intervals during the designated time period.
- Compile the data into a class data table.

(continued on the following page)

INVESTIGATION 12.1 ■ Holding the Heat *(continued)*

USING TECHNOLOGY

1. Load an all-purpose program on your calculator, such as CHEMBIO for Texas Instruments calculators or CASIOLAB for Casio calculators.
2. Use the link cord to firmly attach the calculator to the Data Collector (CBL or EA-100). Turn on both devices.
3. Insert three temperature probes into Channels 1, 2, and 3. Place one probe in each of three aluminum cans.
4. Execute the CHEMBIO or CASIOLAB program on your calculator and follow the screen prompts to enter such information as number of probes, type of probe (thermometers in this case), number of samples and time between samples. If you wish the experiment to run 30 minutes, you should select values that create a total running time of 30 minutes, such as 30 samples 60 seconds apart, or 120 samples 15 seconds apart.
5. When the Data Collector says 'DONE', follow the directions on the calculator screen to transfer and graph the collected data.

Analyzing Your Data and Reporting Your Findings

1. Summarize your results with an appropriate data table, graph, and paragraph. (Hint: See Table 12.2.)
2. Use a graphing calculator or a computerized statistical program to analyze your data using the correlated *t* test or one-way analysis of variance with repeated measures. For options, see the chapter references.
3. Write a report about the experiment (see Chapters 6, 9, 13).

INVESTIGATION 12.2 ▪ Practice Makes Perfect

Question

What effect does practice have on ability to perform a task?

Materials for Each Student

Copy of circled numbers handout
Clock

Safety

If you have difficulty with hand-eye coordination, ask for an alternative role.

Procedure

Read the general procedure outlined below. Plan on a minimum of 30 trials. Draw an experimental design diagram for the experiment and design a class data table for compiling data.

- Keep the handout face down. The handout contains circled numbers from 1 to 59.
- At your teacher's signal, turn the handout over.
- Place your finger on #1, then on #2, and so on until time (30 sec) is called.
- Keep your finger on the last number you reached.
- Record your number in the table below. Turn your handout over.
- Repeat these steps 2 more times for a total of 3 practices.

Number reached in 30 sec

Practice 1	Practice 2	Practice 3
________	________	________

Analyzing Your Data and Reporting Your Findings

1. Summarize your results with an appropriate data table, graph, and paragraph. (Hint: See Table 12.4)
2. Use a graphing calculator or computerized statistical program to analyze your data using the correlated *t* test or one-way analysis of variance with repeated measures. For options, see the chapter references.
3. Write a report about the experiment (see Chapters 6, 9, 13).

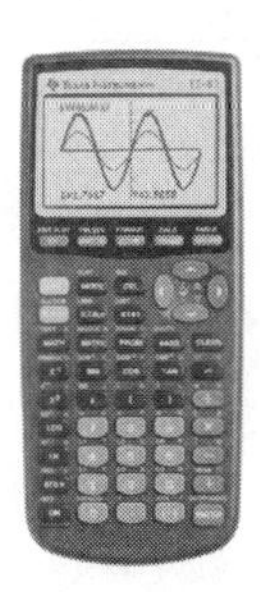

USING TECHNOLOGY

1. In the **STAT** mode of your calculator, enter the numbers reached by each person for the first practice in List 1 and for the last practice in List 2. (See Appendix A, *Using Technology*, for additional help in using the graphing calculator.)
2. Then select **TESTS** and choose the analysis of variance option (ANOVA).
3. Enter List 1 and List 2 as the locations of the data, and calculate.

(continued on the following page)

INVESTIGATION 12.2 ▪ Practice Makes Perfect *(continued)*

18 13 26 41 7 28

38 52 3

8 30 45

49 54 23

36

4 57 33 20 12

11 10 32

47 55

42 51 16 1

39 59

22 19 43 37

35 25

6

15 48 9 29

46 56

2 21

58 24

44 53 34 17

40

27 31 50 14 5

INVESTIGATION 12.3 ▪ Fruit Appeal

Question

What factors affect the oxidation of fruit? How do they interact?

General Materials

- Various fruits such as apples, pears, peaches and nectarines
- Paring knives or scalpels for cutting and peeling fruit
- Ways to heat and cool fruit
- Flat dishes to hold slices of fruit (not metal)
- Bowls or other containers to hold liquids (not metal)
- Cutting boards
- Citrus juices such as lemon and orange
- Commercial products for preserving fruit such as ascorbic acid and "Fruit Fresh"
- Various types of plastic wrap, paper, and zip bags
- Other materials to be determined by the lab groups

General Procedures

As long as an apple has its skin, it retains its color and mineral and vitamin content. Once the skin is removed, however, the fruit comes in contact with oxygen and a chemical reaction, oxidation, takes place. Burning occurs with rapid oxidation while rusting and discoloration occur with slow oxidation.

To prevent oxidation you need to keep oxygen away from the substance; for example, you can smother a fire, paint metal, or use an antioxidant. One type of antioxidant, ascorbic acid, combines with the oxygen and creates a protective colorless covering of oxidized material. Thus, the fruit retains its color, rather than turning brown. Because oxidation is a chemical reaction, it is affected by such factors as temperature, concentration, and surface area.

Use classroom, library, or Internet resources to learn more about the factors that affect discoloration of fruit. Then, identify *two or three variables that you think will interact* to affect the oxidation of fruit. As small groups, or as a class, design an experiment to test your hypotheses.

Draw an experimental design diagram for the experiment, develop a set of procedures, and design a data table for compiling group/class data. (Hint: See Tables 12.5, 12.6, 12.8.)

Conduct the experiment and compile the data.

Analyzing Your Data and Reporting Your Findings

1. Summarize your results with appropriate data tables, graphs, and paragraphs. (Hint: See Tables 12.7 and 12.10.)
2. Use a graphing calculator or computerized statistical program to analyze your data using two- or three-way analysis of variance. For options, see chapter references.
3. Write a report about the experiment (see Chapters 6, 8, 9, 13).

USING TECHNOLOGY

1. Use the **STAT** mode of your calculator to enter the resulting data from your experiment. (See Appendix A, *Using Technology*, for additional help in using the graphing calculator.)
2. Determine the appropriate type of graph for your data and select a scatter plot, histogram, or boxplot to display your data. Graph the data.
3. Using other features of your calculator, perform 1- or 2- variable statistics, predict new values from equations and lines of best fit, or conduct an inferential test as appropriate for the data you collected from your experiment.

INVESTIGATION 12.4 ▪ Collapsing Bridges

Question

How is bridge thickness related to collapsing point?

General Materials

Copier paper, letter size
Scissors
Pennies (all post 1983)
Supports such as books or blocks of wood

Procedure

1. Each lab group should cut 3 sheets of letter size copier paper in half lengthwise to create 6 long strips of paper.
2. Fold up each end of the paper 3 cm (this adds some rigidity to the paper).
3. Make a bridge by supporting 1 strip of paper between 2 books or other objects of similar height. Position the books 15 cm apart so that the paper bridge will span this distance. Center the paper strip over the span with the folded ends pointing upward.

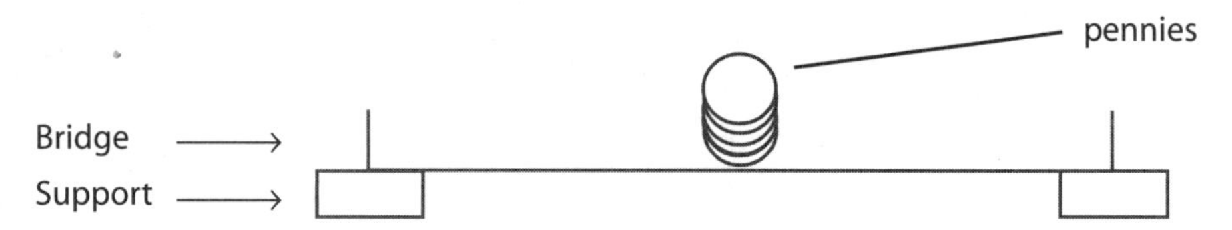

4. Place a penny in the center of the bridge. Continue to stack pennies until the bridge collapses. Record the number of pennies in the table below.
5. Place a second strip of paper under the original strip to form a 2 layer bridge. Repeat Steps 2–4.
6. Repeat Steps 2–5 using a total of 3, 4, 5, and 6 layers of paper.
7. Enter all lab group data into a class data table. Find the mean (average) number of pennies required to collapse the layered bridges.

Lab Group Data: Number of Pennies Required to Collapse Bridges of Various Thickness						
Layers:	1	2	3	4	5	6
Pennies:	______	______	______	______	______	______

Analyzing Your Data and Reporting Your Findings

1. Using only your group's data, make a line graph of number of layers versus the number of pennies. Draw a line-of-best-fit. Use the line to predict the number of pennies required to collapse a bridge made of 7 and 10 layers. Make bridges and test your predictions.

(continued on the following page)

INVESTIGATION 12.4 ▪ Collapsing Bridges *(continued)*

Using the class data, make a line graph. Draw a line-of-best-fit. Use the line to predict the number of pennies required to collapse bridges made of 7 and 10 layers. Make bridges and test your predictions.

In which line graph do you have the greatest confidence? Why?

2. For the class data, determine the slope of the line and the y intercept. Write a mathematical equation to communicate the relationship between the independent variable (number of bridge layers) and the dependent variable (number of pennies). Remember that the general form of a linear equation is $y = ax + b$.

   Use the equation to predict the number of pennies required to break bridges of other thickness, such as 0.5 layers, 2.5 layers, 7.5 layers, and 12.0 layers. How do the values obtained from the equation and from the graph compare?

3. Write a paragraph summarizing your findings about the relationship between the number of bridge layers and collapsing point

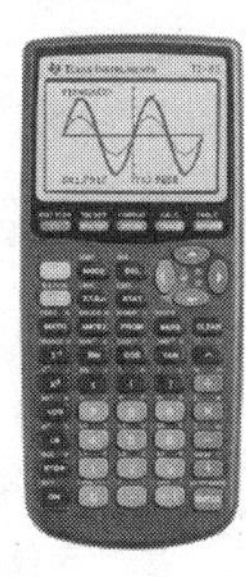

USING TECHNOLOGY

1. In the **STAT** mode of your calculator, enter the number of bridge layers in List 1 and the number of pennies in List 2. (See Appendix A, *Using Technology*, for additional help in using the graphing calculator.)
2. In setting up your graph, select scatterplot as your graph type and List 1 for your x values and List 2 for your y values. Graph the data.
3. Examine the trend of the data and calculate a line-of-best-fit by performing the appropriate regression (equation) analysis. For example, if the general trend of the data appears to be straight, you may wish to calculate a linear regression. Copy the calculated values to an empty Y=, and graph the equation.
4. To predict the number of pennies it would take to collapse a bridge made of layers that were not measured, press **Trace**, then the 'up' arrow and finally the left or right arrow keys to trace along the best fit line to see the predicted y value (number of pennies) for any desired x value (bridge layers). Depending on the brand of your calculator you may have to adjust the 'Window' to predict or extrapolate beyond the minimum or maximum x values.

Extending Your Learning

What other variables do you think affect the strength of bridges? Design experiments to determine the relationship between these variables and breaking point.

Practice

Part I

For each scenario, draw an appropriate experimental design diagram. Some experiments may be represented by more than one design.

1. Larry read that acid rain caused painted surfaces to fade faster. From the literature review, he also found that amount of sunlight and average temperature change also affected paint fading. He established a series of experiments to determine the effect of the three variables on paint fading. Five samples (3 cm x 3 cm) of wood, painted with Brand X Yellow, were exposed to each of the conditions:

   A. pH of acid rain (pH = 2, 4, 6);
   B. ultraviolet light of 5, 10, 15 units;
   C. day-night temperature change of 10°, 20°, 30°C

2. Lou read that bees were attracted to certain colors. She wondered whether crickets also had a color preference. She divided an aquarium into three sections containing three different colored dishes (red, blue, yellow) with mustard seeds (2 g) in each dish. She placed 30 crickets into the aquarium. She observed the number of crickets in each section at the end of 30, 60, 90, and 120 minutes. She recorded the number of grams of mustard seeds consumed at the end of 120 minutes. She was careful to place the aquarium so that the amount of sunlight was the same in all areas.
3. Nitrifying bacteria on the roots of kudzu are partially responsible for the vine's phenomenal growth rate. Because the bacteria are pH sensitive, Julie Neurohr investigated the influence of altered pH on the growth rate of stems, roots, and leaves. Three hundred kudzu cuttings were allowed to establish themselves in soil for 30 days. The width of roots, stems, and leaves were measured prior to planting. The cuttings were subdivided into groups of 50 cuttings; each subgroup was washed with a solution having a pH of 4, 5, 6, 7, 8, or 9. Each subgroup received 1 L of solution every three days and were grown under comparable light and temperature. For six weeks, stem and leaf width were recorded. At the conclusion of the experiment, plant cuttings were removed from the soil and the root width was measured.
4. Vanessa investigated the effect of two different reward schedules on the learning rates of cats. For three weeks, she allowed five cats to walk a maze and recorded the completion time (maximum of 10 minutes). During the next 3 weeks, she rewarded the cats on a regular schedule when they successfully completed the maze. Each cat's completion time was recorded. A three week rest was allowed, after which she rewarded the cats on a random schedule when they walked the maze. Again, times for completion were recorded; Brand X cat treats were used as a reward. The maze, brand of treat, and interval since the last feeding remained the same.
5. Mary's family encountered conflicting evidence on the effectiveness of various brands of water softening units. Before purchase, Mary investigated the effectiveness of Brands X, Y, and Z with water prepared to three levels of hardness (50 mg/L, 100 mg/L, and 150 mg/L). The hardness, hardness reduction, and increase in sodium concentration after treatment were determined. Ten samples of each type of water were exposed to each softener.

(continued on the following page)

Practice *(continued)*

6. Wildlife biologists typically determine weight of deer with scales. In field conditions, a simpler technique is desirable. Kelly Raybourne investigated the feasibility of chest girth measurements as a substitute. From authorized big game checking stations east and west of the Blue Ridge Mountains, data were secured on sex, age, weight, and chest girth. Across all subpopulations, the relationship between chest girth and weight was investigated. The relationship between chest girth and weight for each subgroup of deer (male and female, young and old, eastern and western Blue Ridge) was also analyzed.
7. Courtney Lyon knew that fish were adversely affected by thermal, phosphate, and acidic pollution. In general, each of these factors negatively influenced respiration, as indicated by a heightened respiratory rate (gill beats/minute). In the experiment, Courtney measured the change in gill beats/minute after 12 hours of exposure to various conditions. In the experiment, she used 120 fish divided into 24 groups, each containing 5 fish. She used three different independent variables. For the first pH, she used three levels (6.5, 5.5, 4.5). For temperature, she used 10°C, 20°C, 30°C, and 40°C. Phosphate concentrations were 0 and 3 ppm. Draw an experimental design diagram and describe the main effects and interactions she could explore.

Part II

Using techniques described in this chapter, determine whether the research hypotheses given in the following scenario are supported by the data.

TABLE 12.12 Data for Weed Killers and Grass Concentration

	Time				
Concentration	8:00 A.M.	10:00 A.M.	12:00 P.M.	2:00 P.M.	4:00 P.M.
40%	20	19	10	10	15
	19	18	8	9	17
	20	16	9	11	16
	20	17	11	9	14
	20	16	10	10	15
20%	19	13	4	5	10
	18	12	5	6	9
	16	14	6	7	11
	17	11	4	6	8
	15	12	4	5	7
10%	18	8	3	7	9
	15	7	2	6	7
	16	6	3	7	8
	15	9	4	5	9
	14	8	2	6	6

(continued on the following page)

Practice *(continued)*

1. John read that the biorhythms of organisms influence metabolism of drugs; thus, side effects of drugs can be minimized by prescribing lower concentrations at peak times. Generally, Brand X weed killer is more effective in higher concentrations and when applied early in the day. Because the weed killer acts by interfering with water uptake during photosynthesis, John hypothesized that concentration and time of application would interact to determine maximum effectiveness. John subdivided a field into 15 plots, each containing 5 grass plants. He applied 10 percent, 20 percent and 40 percent X (recommended) at 5 times (8:00 A.M., 10:00 A.M., 12:00 P.M., 2:00 P.M., and 4:00 P.M.). After two weeks, the number of dead plants per plot were recorded (see Table 12.12).

CHAPTER 13

Preparing Formal Papers

Objectives

- State reasons why a review of the literature should be conducted before you begin an independent research project.
- Describe important information to include in a review of the literature.
- Write a formal review of the literature for an independent research project.
- Identify the elements of a scientific research paper—title page, abstract, introduction, methods and materials, results, discussion-conclusion, and bibliography.
- Write a scientific research paper for an independent science project.
- Use a checklist to evaluate a scientific research paper and to identify needed improvements.

National Standards Connections

- Formulate and revise scientific explanations and models using loci and evidence (NSES).
- Recognize and analyze alternative explanations and models (NSES).
- Design a statistical experiment to study a problem, conduct the experiment, and interpret and communicate the outcomes (NCTM).

Scientists write formal papers to establish the rationale for a proposed research study, as well as other papers to report the findings of the research study. The rationale for the study is established in a review of the literature that includes important background information on the variables, an analysis of prior research on the topic, and a statement of the research problem. Your research design and understanding of underlying scientific principles will be improved by reviewing the literature prior to finalizing the experimental design, procedures, and collection of data.

Scientists typically report research through concise research papers. The format of these papers, which is seen in articles in scientific journals, includes a title, abstract, introduction, methods-materials, results, discussion-conclusion, and bibliography. Most science competitions require that you report your research in this format. In reporting your research, you will use information from the literature review to write the introduction, to discuss the results, and to formulate a conclusion. In this chapter, procedures for writing a review of the literature and a scientific research paper are summarized and an example of a student research paper is provided.

REVIEW OF THE LITERATURE

Reviews of the literature typically consist of three major components—general background information, an analysis of prior research, and a problem statement. Other components include the title page and a bibliography. Because scientists do not consistently use a single style manual, no particular manual is recommended. Generally, you should use the format required by your school's English department. If science competitive events in your area require a particular style manual, however, you will need to use that manual for both the literature review and the science research paper.

General Background Information

In this component, you should report information on the independent and dependent variables, the specific subject of the study, and any specialized procedures. Techniques for helping you with initial library research, including the documentation of sources and the preparation of note cards, are described in Chapter 7, *Using Library Resources*. Structured guidelines to help you determine the essential information to collect and to provide an outline for reporting the literature review are found in Table 13.1 *How to Write a Review of the Literature.*

Begin by broadly categorizing the variables in your research project as plant, animal, protist, matter, energy, process/procedures, or behavior. Remember the traditional scientific disciplines—plant (botany), animal (zoology), matter (chemistry), energy (physics), and earth's processes (geology, meteorology, and so on), and human behavior (psychology and sociology). Knowledge of discipline names will assist you in locating the right information. These categories include the majority of variables and subjects investigated by beginning researchers. Look at the following sample titles of research papers. Do you agree with the subject areas represented by the independent and dependent variables? Which subject areas would you use to describe the variables in your research project?

Title: The Effect of Aspirin on the Rooting of Kudzu Vines
IV: Aspirin (Matter—Chemistry)
DV: Rooting of Kudzu Vines (Plant—Botany)

Title: The Effect of Rotation on the Crystallization of Zinc, Copper, and Lead Sulfate
IV: Rotation (Energy—Physics)
DV: Crystallization of Zinc, Copper, and Lead Sulfate (Matter—Chemistry)

Title: The Effect of Three Methods of Terracing on Erosion of Field Soils
IV: Methods of Terracing (Process/Procedure—Earth Science)
DV: Erosion of Field Soils (Matter—Earth Science)

Title: The Effect of Magnetic Fields on Budding of Yeast
IV: Magnetic Field (Energy—Physics)
DV: Budding of Yeast (Protist—Botany)

Title: The Effect of Weekly Stress on the Aggressive Behavior of Toll Attendants
IV: Weekly Stress (Behavior—Psychology)
DV: Aggressive Behavior of Toll Attendants (Behavior—Psychology)

The broad categories into which your research variables fall determine the critical information that you should find and report.

For the Independent Variable, Aspirin, Form of Matter: Briefly describe the substance's chemical and/or common name, formula, physical properties, chemical properties, method of production, and uses. Provide detailed information on its characteristics—chemical and physical properties—that are directly applicable to the study. If you have more than one form of matter, such as aspirin and vinegar, provide this information for each; then, describe similarities and difference of the substances. Based upon the information, predict how the forms of matter will act in the experiment.

For the Dependent Variable, Rooting of Kudzu Vine, Plant: Provide the plant's common name, scientific name, and classification. Briefly describe the plant's (a) habitat; (b) anatomy including root, stem, leaf, and flower; (c) physiology including how it obtains needed materials, moves, eliminates waste, responds to stimuli, and so on; (d) life cycle; and (e) behaviors or responses to the environment. Provide detailed information in areas that are directly applicable to the study, such as structure and growth of roots. If you have more than one type of plant, such as kudzu and sweet potato vines, provide information on each; then, describe similarities and differences of the plants. Based on the information, predict how the plants will act in the experiment.

Using information obtained through the above process, you can write the general background portion of the literature review. Brief descriptions of critical information to include about forms of energy, animals, matter protists, processes, behavior, and process/procedures, are included in Table 13.1.

Analysis of Prior Research

In this component of the literature review, you should describe related research studies. You should also analyze the studies and establish the implications for your proposed research projects: Although this component is the heart of a professional scientist's literature review, you may be limited by the extent of your scientific knowledge and by the resources of local libraries.

Techniques for assisting you to take notes from scientific journals and abstracts are described in Chapter 7. Critical information to be recorded about each research study includes purpose/hypothesis, experimental design, brief description of procedures, major findings, conclusion, and recommendations for further research. You can write a brief related description of research studies from such notes.

Because analyzing and evaluating critical information from several research studies is more difficult, begin by listing areas of agreement and disagreement among the studies. Propose an explanation for any differences among the findings of the studies. Offer explanations for variations in the materials or procedures used in the studies. For example, you might ask: "Were the same types of plants used in the studies?", "What did you learn about aspirin production in your background reading that might explain differing results with various brands of aspirin?" When you have identified differences, ask what research studies could be conducted to resolve conflicting findings or to address unanswered questions. Describe how your proposed study will address these areas, thus expanding an understanding of the topic. When you have successfully identified similarities, differences, and potential research questions, write a paragraph that summarizes your analysis.

Statement of the Problem

The literature review ends with a brief summary of the purpose and rationale for the study being undertaken and the research hypotheses (see Table 13.1). In the simple scientific reports described in Chapter 6, the problem statement

TABLE 13.1 How to Write a Review of the Literature

Before you begin the literature review, identify the independent and dependent variables for your research study and follow your teacher's instructions for classifying the variables into the following broad categories.

Animal	Behavior	Matter	Process/Procedure
	Energy	Plant	Protists

Title: Write a sentence that identifies the independent and dependent variables for the investigation.

General Background Information: Describe important characteristics of the independent and dependent variables. Use the following descriptions that relate to your study to help focus your library research and to provide an outline for summarizing the background information.

Animal: Provide the animal's common name, scientific name, and classification. Briefly describe the animal's (a) habitat; (b) anatomy including information on the skeletal, muscular, circulatory, nervous, digestive, and excretory systems; (c) physiology including how it obtains needed materials, moves, eliminates wastes, responds to stimuli, and so on; (d) life cycle; and (e) behaviors or responses to the environment. Provide detailed information in areas that are directly applicable to the study. If you have more than one animal, provide this information on each animal; then, describe similarities and differences of the animals. Based on the information, predict how you think the animals will act in your experiment.

Behavior: Briefly describe the type of behavior, the primary factors influencing the behavior, the value of the behavior, and methods for describing the behavior. Identify critical factors to consider in selecting the sample for the study and for designing the procedure. If you have more than one type of behavior, provide this information on each type of behavior; then, describe similarities and differences. Based on the information, predict how you think the subjects will behave in your experiment.

Energy: Briefly describe the form of energy including how it is produced, measured, transformed into other kinds of energy, and interacts with matter. Cite common examples of the form of energy found in nature or produced by humans. If you have more than one form of energy, provide this information on each form; then, describe similarities and differences of the energy forms. Based on the information, predict how you think the forms of energy will act in your experiment.

Matter: Briefly describe the substance's chemical and common names, formula, physical properties, chemical properties, method of production, and uses. Provide detailed information on characteristics that are directly applicable to the study. If you have more than one form of matter, provide this information on each substance; then, describe similarities and differences of the substances. Based on the information, predict how you think the forms of matter will act in your experiment.

Plant: Provide the plant's common name, scientific name, and classification. Briefly describe the plant's (a) habitat; (b) anatomy including root, stem, leaf, and flower; (c) physiology including how it obtains needed materials, moves, eliminates waste, responds to stimuli, and so on; (d) life cycle; and (e) behavior or response to the environment. Provide detailed information in areas that are directly applicable to your study. If you have more than one type of plant, provide this information on each; then, describe similarities and differences of the plants. Based on the information, predict how you think the plants will act in your experiment.

Process/Procedure: Describe the purpose of the process, the major steps, where it occurs, and how it relates to the experiment. If more than one process is involved, provide this information on each process; then, describe similarities and differences of the processes. You may also include pros and cons of each process. Based on the information, predict how you think the processes will affect your experiment.

Protists: Provide the protist's common name, scientific name, and classification. Briefly describe the protist's (a) habitat; (b) anatomy; (c) physiology including how it obtains needed materials, moves, eliminates waste, responds to stimuli, and so on; (d) life cycle; and (e) behavior or response to the envi-

(continued on the following page)

TABLE 13.1 (continued)

ronment. Provide detailed information in areas that are directly applicable to the study. If you have more than one type of protist, provide this information on each; then, describe similarities and differences of the protists. Based on the information, predict how you think the protists will act in your experiment.

Analysis of Prior Research (Optional): Briefly summarize scientific research studies directly related to your study. Include the purpose, procedures, major findings, and recommendations for further study. If you review more than one study, describe similarities and differences among the research studies. Suggest research studies that need to be conducted to resolve differences or to address unanswered questions.

Statement of the Problem: Describe the rationale, purpose, and hypotheses for the investigation. Use three questions to guide your writing of the introduction:

- Why will you conduct the experiment? (Rationale)
- What do you hope to learn? (Purpose)
- What do you think will happen? (Hypothesis)

Bibliography: List all books, papers, journal articles, and communications cited in the paper. Follow the prescribed bibliographic style manual precisely.

Special Instructions: Follow the teacher's instructions for the length of the paper, format of the paper, style manual, and deadline for submission.

served as the introduction for the report. As with simple reports, you can use three questions to guide your writing of the problem statement.

- Why will you conduct the experiment? (Rationale)
- What do you hope to learn? (Purpose)
- What do you think will happen? (Hypothesis)

FORMAL RESEARCH PAPERS

Rules vary with the competitive event. If you plan to enter a research paper in a competitive event, you must know and comply with the rules of the competition. Because papers that do not comply will almost always be disqualified, it is critical that you follow the specified guidelines. Most competitive guidelines specify an absolute maximum length. Regulations stating the size of the paper, spacing, margins, and the number of characters per inch are also very precise and will vary. Because looks count, print the final version of the paper on a letter-quality printer. Printing by dot matrix printers should be avoided and, in fact, is sometimes prohibited. Subject to specific guidelines, most formal scientific research papers include a title, abstract, introduction, methods and materials, results, and a discussion-conclusion. Other components include the title page, bibliography, acknowledgements, and appendix. General guidelines for writing these components are summarized in Table 13.2 *How to Write a Scientific Research Paper*. Modify these guidelines as needed to comply with your school's or the competition's requirements.

Reading student research papers from past years and studying the components outlined in Table 13.2 will help you write your own paper. A high school junior's paper on the effect of gibberellic acid on the closing speed of the Venus flytrap is provided to help you get started. Remember, at this point you already know how to write three of the major report components—the methods and materials, results, and discussion-conclusion—you also have the essential information to write an introduction from your review of the literature. You can then easily add minor components, such as the title page, abstract, bibliography, acknowledgements, and appendices (see Chapters 4, 6, and 9).

TABLE 13.2 How to Write a Scientific Research Paper

Title Page
Include the name of the project, your name, the teacher's name, the class, and the date. Follow guidelines provided by your teacher for the format of the page and for additional information required for the competition, for example, school's and parents' names and addresses, hours spent on project, and category.

Abstract
Write a concise summary of your project that includes the problem, hypothesis, procedures, principal results, and conclusions. Do not exceed 250 words.

Introduction
From the review of the literature, summarize information essential for understanding the research project. Include only critical background information on the independent and dependent variables and research studies that directly relate to the research problem. Establish a strong rationale for the study by emphasizing unresolved issues or questions. Conclude by stating the purpose of the study and the research hypotheses.

Methods and Materials
In paragraph form, describe the materials and procedures used to conduct the study. Step listings are not acceptable. Provide sufficient detail to allow a reader to repeat the study. Include precise descriptions of the sample, any apparatus that was constructed or modified for the study, and methods of data collection.

Results
Present the data collected in the experiment in tables and graphs; summarize the data in narrative form. Include statistical analysis of the data. Do not include raw data; if necessary, the raw data can be placed in the appendix. Include only information collected during the study.

Discussion-Conclusion
Restate the purpose of the study, the major findings, and support of the hypothesis by the data. Focus on interpretation of the results. Compare findings with other research; propose explanations for discrepancies. Be sure to provide appropriate literature citations. In addition, make suggestions for procedural improvements and recommendations for further study.

Bibliography
List all books, papers, journal articles, and communications cited in the paper. Follow the prescribed bibliographic style manual precisely.

Acknowledgements
Credit assistance received from mentors, parents, teachers, and other sources. As directed by your teacher, include statements by mentors that certify the precise nature of your work. Forms may also be required for human or other vertebrate experimentation.

Appendix
Include critical information that is too lengthy for the main section of the paper, such as raw data, additional tables and graphs, copies of surveys or tests, and diagrams of specialized equipment.

Special Instructions
Precisely follow your teacher's or the competition's guidelines for maximum length of the paper, typing format, use of color or photographs, deadline for submissions, and style format. With competitions, failure to comply may mean automatic rejection.

STUDENT RESEARCH PAPER

The Effect of Gibberellic Acid on the Closing Speed of the Venus Flytrap

by
Kimberly P. Bryant
Patrick Henry High School, Ashland, VA

Abstract

The Venus flytrap is a small carnivorous plant possessing bilobed leaves in which it captures its prey. The closing of these leaves around a suitably-sized nitrogen-containing object takes place in two distinct phases, one in which the lobes of the leaf snap shut, trapping the prey, and one in which the lobes slowly squeeze together to facilitate digestion. It is believed that the Phase I reaction is brought about by changes in turgor pressure after a trigger hair in the leaf is stimulated, but that the Phase II reaction is an actual growth response. This experiment was designed to test the hypothesis that treating the Venus flytrap with gibberellic acid, a growth hormone, would increase the speed of the Phase II reaction. Ten Venus flytraps were watered with 1 ppm gibberellic acid solution, while ten plants were watered with distilled water every other day for ten days; at the end of this time, the speed of both closing reactions was tested using five leaves from each plant. Though the mean time of both phases was shorter in the experimental group, the results of the *t* tests showed that only the data from Phase II were significant at the 0.05 level of significance. The research hypothesis that the Phase II closing speed of the Venus flytrap would be increased by the application of gibberellic acid was therefore supported.

Introduction

The Venus flytrap, *Dionaea muscipula,* a member of the *Droseraceae* family, is a carnivorous plant found in damp areas in the eastern part of North Carolina. The roots are small and are used mainly for the absorption of water by the plant; the Venus flytrap obtains most of its nutrients from the insects it traps in its bilobed leaves. Three small triangularly placed filaments are on the inner surface of each lobe of the leaf, perpendicular to the main surface but capable of bending over. When these filaments are stimulated, the lobes of the leaf close together, leaving a small pocket-like enclosure between them. Here the plant secretes digestive enzymes that consume insects or other nitrogen-containing matter. Leaves that have been stimulated to close but that contain no nitrogenous matter generally open within 24 hours, but leaves that close over nitrogenous matter may not open for several days. After two or three digestions the leaf usually dies or is too sluggish to trap any more prey, but leaves may open and close many times if they fail to trap any digestible matter (Darwin, 1972).

Carnivorous plants were for the most part dismissed as exaggerated traveller's tales until Charles Darwin published *Insectivorous Plants* in 1874; this book sparked interest in the phenomenon and gave rise to many experiments on Venus flytraps. Darwin himself performed several experiments by which he determined that the plant is a discerning gourmand, rejecting bits of wood or wax or thick-shelled beetles in favor of ants and spiders. He also determined that the spikes at the margins of the leaves allow very small insects to escape but trap larger insects, saving the plant's powers of digestion for worthwhile prey (Darwin, 1972). One Dr. Curtis in the late nineteenth century in North Carolina discovered that the plants could be successfully maintained on a diet of beef, but that high-protein cheese seemed to cause indigestion and resulted in the death of the leaf (Emboden, 1974). J.M. MacFarlane found in 1872 that two stimuli are required for the closing of the trap; either the same filament must be touched twice or two filaments must be touched (Lloyd, 1976). Darwin had previously believed that only one stimulus was necessary but had noticed that a very light stimulus might not provoke the closing of the leaf (Darwin, 1972).

F.J.F. Meyen made the first attempt to explain the mechanism by which the Venus flytrap closes in 1839, but his explanation that the leaves closed by way of a spring-like mechanism was inadequate. Later researchers hypothesized that the turgor of cells in the leaf must be a factor in closing. Darwin noted that closure appeared to take place in two parts; in the first, the leaf responds quickly to close itself approximately, and in the second, the leaf's lobes press slowly together and the outer surface of the lobes expands (Lloyd, 1976). J. Burdon-Sanderson believed that changes in turgor were the only factors in this reaction, but in 1877, A. Batalin's experiments suggested that actual growth takes place in the leaf during the process of closing and digestion (Lloyd, 1976). It is now thought that the second phase of closing, called the narrowing phase, is caused by a growth spurt; the outer surface grows and forces the lobes inwards, and when the leaf reopens the inner surface grows, forcing the lobes outwards (Slack, 1979).

The gibberellins were discovered in 1926 by E. Kurosawa, a Japanese plant pathologist studying *bakanae,* a disease in which rice seedlings become tall and spindly. It was soon established that the disease was caused by fungus *Gibberella fujikuroi,* and in 1935 T. Yabuta named the active factor in cultures of *G. fujikuroi* "gibberellin." This research was unavailable to the Western world until about 1950, when American and British research teams began trying to purify gibberellin from filtrates of the fungus. In 1954–55 a new compound was isolated from the filtrate, and it is now known as gibberellic acid (GA), $C_6H_{22}O_6$. Since then it has been established that gibberellins occur naturally in a great many species of plants, and at least 52 different gibberellins have been isolated, all having the same basic chemical skeleton (Moore, 1979).

GA works with other plant growth hormones such as the auxins to control cell elongation. The application of GA can even induce normal growth in genetically dwarfed plants. It is also an active factor in such plant growth systems as fruit-set, bolting, and flowering (Devlin, 1975). Although a great deal of experimentation has been conducted using GA, the way in which this growth hormone works upon the plant is not yet fully understood.

Because it is thought that the second phase of trap closing in Venus flytraps is a growth response and gibberellic acid has been shown to induce growth, it is possible that the application of GA may affect the closing speed of the Venus flytrap in some way, perhaps increasing the speed of the second phase of the closing reaction. The purpose of this experiment, then, is to ascertain whether GA affects the closing speed of Venus flytraps. It is unlikely that the results of this experiment, whatever they be, will have any lasting benefits for humanity besides, perhaps, the satisfaction of curiosity; but, after all, one of the primary purposes of science itself is the satisfaction of human curiosity.

Materials and Procedures

The materials used for this experiment were 20 Venus flytraps, a plant light, 1.5 liters of 1 ppm gibberellic acid solution, 1.5 liters of distilled water, a pair of rubber gloves, a 100-milliliter graduated cylinder, tweezers, a stopwatch, and shredded roasted turkey.

The Venus flytraps were set up under the plant light in two groups: the control and experimental groups. On every other day for 10 days, half of the plants were watered with 15 milliliters of distilled water and half of the plants were watered with 15 milliliters of gibberellic acid solution. All testing was done on the eleventh day. The position and reactions of 5 leaves on each plant were recorded; a small piece of shredded turkey was placed in each leaf before the trigger hairs were stimulated, and the Phase I closing time and speed were recorded for each leaf. While Phase I experimentation continued, frequent observations for signs of Phase II closing were made. The time of complete Phase II closing for each leaf was also recorded. The mean time for each closing phase was calculated for both the experimental and control groups, and *t* tests were performed on each set of data at the 0.05 level of significance.

Results

Data Table, Graph, and Statistical Table A show the effect of 1 ppm solution of gibberellic acid on the Phase I closing speed of Venus flytraps. The mean Phase I closing speed of the plants to which gibberellic acid was added (1.7 sec) was less than that of the control group (1.8 sec). There was a greater variation among the control group (range = 2.7 sec) than there was among the experimental group (range = 1.9 sec). Substantial variation existed in both the control group (standard deviation = 0.48) and in the experimental group (standard deviation = 0.46). A *t* test was used to test the null hypothesis of no significant difference at the 0.05 level of significance. The calculated *t* value (t = .83) was not significant at the 0.05 level of significance, and the null hypothesis, stating that no significant difference would exist between the groups, was therefore accepted.

DATA TABLE A
Effect of GA on Phase I Closing Speed (sec)

Type of Descriptive Data	Control 0 ppm GA	Experimental 1 ppm GA
Mean	1.8 sec	1.7 sec
Minimum	1.1	0.9
Maximum	3.8	2.8
Range	2.7	1.9
Number	50	50

GRAPH A
Effect of GA on Phase I Closing Speed (sec)

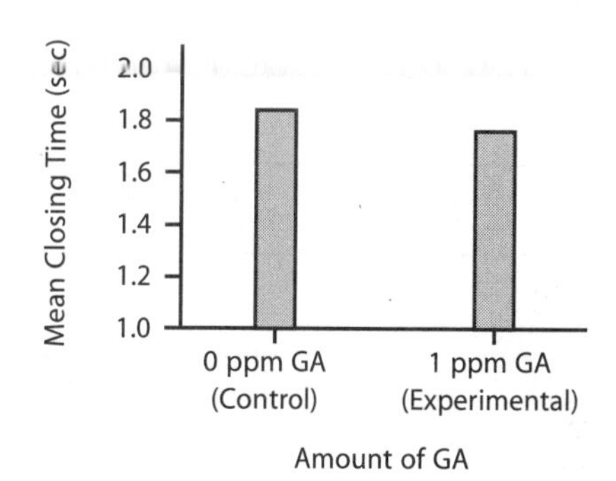

STATISTICAL TABLE A
Effect of GA on Phase I Closing Speed (sec)

Mean difference	0.10 sec
Control standard deviation	0.48
Experimental standard deviation	0.46
Significance level	0.05
Standard *t*-value	2.01
Obtained *t*-value	0.83
Not Significant	

Data Table, Graph, and Statistical Table B show the effect of 1 ppm solution of gibberellic acid on the Phase II closing speed of Venus flytraps. The mean Phase II closing speed of the plants to which gibberellic acid was added (90.2 min) was less than that of the control group (101.2 min). There was a greater variation among the control group (range = 218.0 min) than there was among the experimental group (range = 167.0 min). Substantial variation existed in both the control group (standard deviation = 55.5) and in the experimental group (standard deviation = 46.5). A *t* test was used to test the null hypothesis of no significant difference at the 0.05 level of significance. The calculated *t* value ($t = 7.68$) was significant at the 0.05 level of significance, and the null hypothesis, stating that no significant difference would exist between the groups, was therefore rejected.

DATA TABLE B
Effect of GA on Phase II Closing Speed

Type of Descriptive Data	Control 0 ppm GA	Experimental 1 ppm GA
Mean	101.2 min	90.2 min
Minimum	27.0	21.0
Maximum	245.0	188.0
Range	218.0	167.0
Number	50	50

GRAPH B
Effect of GA on Phase II Closing Speed

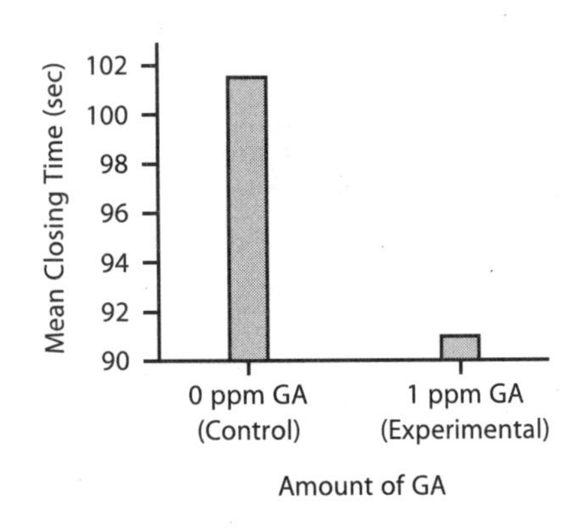

STATISTICAL TABLE B
Effect of GA on Phase II Closing Speed

Mean difference	11.0 min
Control standard deviation	55.5
Experimental standard deviation	46.5
Significance level	0.05
Standard *t*-value	2.01
Obtained *t*-value	7.68
Significant	

Conclusion

This experiment was designed to test the effects of a 1 ppm solution of gibberellic acid on the speed of both closing phases of the Venus flytrap's leaves. It was hypothesized that because the second closing phase is thought to be a growth response, the application of gibberellic acid would increase the speed of the reaction; however, the first phase of closing, which is basically a change in turgor pressure, would not be affected. It was found that the mean time for each phase was less in the group of plants treated with gibberellic acid. These data appeared not to support the hypothesis concerning Phase I but to support the hypothesis concerning Phase II; that is, the data suggested that the speed of both reactions was increased by the application of gibberellic acid.

However, results of *t* tests completed on the data showed that there was a significant difference between the control and experimental groups only with the second phase of closure; the research hypothesis was supported by the data. The first phase of closure was unaffected by gibberellic acid, whereas the speed of the second phase of closure was increased. The data may have been affected somewhat by uncontrolled environmental variables such as temperature and humidity; this experiment could have been improved by keeping the Venus flytraps in a climatorium, where more outside variables could be controlled. Using more plants would also increase the accuracy of the data. For further research, extensive records could be kept on each leaf so that individual variations in closing speed could be correlated and accounted for; then the effect of gibberellic acid on the Venus flytrap could be established more firmly.

ORIGINAL REFERENCES CITED

Acid explanation of Venus flytrap spring. *Science News*, January 15, 1983, p. 41.
Acid flux triggers the Venus flytrap. *New Scientist*, March 3, 1983, p. 582.
The Allure of Carnivorous Plants. *New Scientist*, February 6, 1986, p. 32.
Baker, N.R., Davies, W.J., & Ong, C.K., editors. *Control of Leaf Growth.* Cambridge: Cambridge University Press, 1985.
Boxer, Sarah. The subtlest assassins. *The Sciences*, May/June 1983, pp. 7–8.
Cell growth causes plant to shut its trap. *New Scientist*, January 8, 1981, p. 72.
Darwin, Charles. *Insectivorous plants.* New York: AMS Press Inc., 1972.
Davies, P.J., editor. *Plant hormones and their role in plant growth and development.* Dordrecht, The Netherlands: Martinus Nijhoff Publishers, 1987.
Devlin, Robert M. *Plant physiology.* New York: Litton Educational Publishing, 1975.
Emboden, William A. *Bizzare plants: Magical, monstrous, mythical.* New York: Macmillan Publishing Company, Inc., 1974.
Lloyd, Francis Ernest. *The carnivorous plants.* New York: Dover Publications, Inc., 1976.
Moore, Thomas C. *Biochemistry and physiology of plant hormones.* New York: Springer-Verlag New York Inc, 1979.
Scott, Tom K. *Hormonal regulation of development II: The functions of hormones from the level of the cell to the whole plant.* Wurzburg, Germany: Springer-Verlag Berlin Heidelburg, 1984.
Slack, Adrian. *Carnivorous plants.* Cambridge, Massachusetts: The MIT Press, 1979.
Wallace, Robert A., King, Jack L., & Sanders, Gerald P. *Biology: The science of life.* Glenview, IL: Scott, Foresman, and Company, 1986.
Williams, Stephen E., Bennett, Alan B. Leaf closure in the Venus flytrap: An acid growth response. *Science*, December 10, 1982, pp. 1120–22.

ACKNOWLEDGEMENTS

I would like to thank Mrs. Eleanor Tenney, who gave me the idea for this project, Mrs. Pamela Gentry, who helped me put it together, and Dr. Julia Cothron, who explained the statistics to me. I would also like to thank my mother, who drove for miles to find the best deal in Venus flytraps and then patiently answered the questions of all the people who wanted to know why anyone would want so many strange-looking plants, and my father, whose first-rate skills at on-the-spot school-project carpentry provided a home for the Venus flytraps.

USING TECHNOLOGY
SURFING THE WEB FOR CARNIVOROUS PLANTS

More than a decade ago, Kimberly Bryant conducted her study of the effect of Gibberellic Acid on the closing speed of the Venus flytrap. Then, middle and high school students had limited access to scientific references, unless a college/university library was nearby. Obtaining up-to-date references was much more difficult than today, for the Internet has revolutionized your abilities to obtain information.

If you were interested in carnivorous plants, such as the Venus flytrap, they would need to use much more recent references than the ones cited in Kimberly Bryant's paper. Except for classical studies, such as the ones by Darwin and Kurosawa, scientists generally do not cite references that are more than ten years old. Fortunately, you have search engines and the electronic world at your fingertips. Here are some handy websites we found on the topic:

- Botany Encyclopedia of Plants (www.botany.com)
- Botany Libraries—Harvard University Herbaria (www.herbaria.harvard.edu/Libraries/libraries)
- Botany Smithsonian Databases (http://nmnhwww.si.edu/botany/database.htm)
- Botanical Society of America (www.botany.org)
- Carnivorous Plants: Cultivation (www.geocities.com/RainForest/Vines)
- The Carnivorous Plant (www.sarracenia.com)
- International Carnivorous Plant Society, Inc. (www.carnivorousplants.org)
- Internet Directory for Botany (www.botany.net)

Check out the sites we have identified. Then, use your ingenuity to find others. Did you learn more about the effect of gibberellic acid on Venus flytraps? Did you find other interesting questions to investigate?

EVALUATING YOUR SKILLS

After you have completed a draft of your research paper, check it using the criteria listed in Table 13.3 *Checklist for Evaluating a Scientific Research Paper*. Form a study group with your classmates to analyze each others' scientific research papers and revise as needed. Remember, the more suggestions you have to improve your paper, the more successful you will be with your class assignment or in a scientific competition.

TABLE 13.3 Checklist for Evaluating a Scientific Research Paper

Criteria	Self	Peer/ Family	Teacher
Introduction			
Background on IV and DV			
Review of prior research			
Statement of problem			
Methods and Materials			
All materials/equipment included			
Clear/precise description			
Results			
Data tables			
Graphs			
Paragraphs of results			
Statistical Test—optional			
Discussion and Conclusion			
Purpose of experiment			
Major findings			
Support of research hypothesis by data			
Comparison with other research			
Explanations for findings			
Recommendations			
Writing			
Logical organization/effective transitions			
Sentence/paragraph structure			
Grammar/spelling			
General Format			
Abstract			
Title page			
Footnotes			
Bibliography			
Acknowledgments			
Appendix			

REFERENCES

Clyne, D. (1998). *Plants of prey*. Wisconsis, WI: Gareth Stevens.

D'Amato, P. (1998). *The savage garden: Cultivating carnivorous plants*. Berkeley, CA: Ten Speed Press.

Gibaldi, J. (1998). *MLA style manual and guide to scholarly publishing* (2nd ed.). New York: Modern Language Association of American.

Bryant, K.P. *The effect of gibberellic acid on the closing speed of the Venus fly-trap*. Paper presented at the meeting of the Virginia Junior Academy of Science, First Place in Botany Division and Botany Award, Richmond, VA: Virginia Commonwealth University.

Chicago Editorial Staff. (1993). *The Chicago manual of style: The essential guide for writers, editors, and publishers* (14th ed.). Chicago: University of Chicago Press.

Gentle, V. (1996). *Bladderworts: Greasy cups of death (Bloodthirsty Plants)*. Milwaukee, WI: Gareth Stevens.

Gentle, V. (1996). *Pitcher plants: slippery pits of no escape (Bloodthirsty Plants)*. Milwaukee, WI: Gareth Stevens.

Gentle, V. (1996). *Venus Fly Traps and waterwheels: Spring traps of the plant world (Bloodthirsty Plants)*. Milwaukee, WI: Gareth Stevens.

Juniper, B.E., Robins, R.J., & Joel, D.M. (1989). *The carnivorous plants.* Orlando, FL: Academic Press.

Publication manual of the American Psychological Association (4th ed.). (1994). Hyattsville, MD: American Psychological Association.

Simons, P. (1996. August). When a carnivore is not a carnivore. *New Scientist*, 151, p. 16.

Turabian, K. L. (1996). *A manual for writers of term papers, theses, and dissertations* (6th ed.). Chicago: University of Chicago Press.

Related Web Sites

http://www.dade.k12.fl.us/us1/science/prod03.htm

http://www.mcrel.org/resources/links/index.asp

http://www.eduzone.com/Tips/science/SHOWTIP2.HTM (report section)

CHAPTER 14

Presenting Scientific Research

Objectives

- Identify important features of good oral presentations.
- Make an oral presentation on an independent research project.
- Identify important features of good displays for science fairs and poster sessions.
- Make a science fair or poster display on an independent research project.
- Use a typical science project evaluation form to evaluate presentations of scientific research and identify needed improvements.

National Standards Connections

- Communicate and defend a scientific argument (NSES).
- Think critically and logically to make relationships between evidence and explanations (NSES).

Unlike writing your scientific paper, which must follow a standard format, you have several ways you could present your research paper. These include oral reports, poster sessions, and displays. The method of presentation differs with each competitive event. Science fairs, the most common competitive event at local and regional levels, typically require a three-panel display board. Poster sessions held by the American Academy of Science provide a low-cost alternative to science fair displays. Competitive events sponsored by industries, special interest societies, academies of science, and government agencies require oral presentations of projects previously selected from written reports. Typically, competitive events that require displays or posters also require written reports and oral presentations of projects to judges. This chapter describes the general format for each type of presentation.

SCIENCE FAIR DISPLAYS

Most local science fairs follow the requirements of the International Science and Engineering Fair for exhibit size and components. Exhibits cannot exceed a depth of 76 cm (30 in), width of 122 cm (48 in) and height of 274 cm (9 ft) including the height of the table. Guidelines prohibit displays

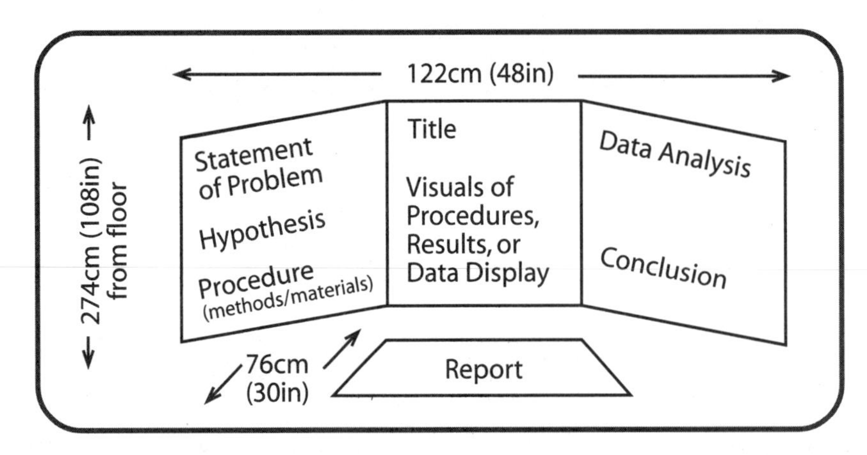

Figure 14.1 General Diagram of Science Fair Display.

of vertebrate and invertebrate animals; photographs of surgical techniques such as autopsies; exhibition of most human parts; Class III or IV lasers; and fuels, foods, microbes, and chemicals that endanger public safety. Because of state quarantine requirements, displays of plant or agricultural products are discouraged. Consult the rules of the International Science and Engineering Fair, published yearly, for specific requirements. The format of a science fair display board is similar to the components of a simple science report: title, statement of problem, procedure (methods-materials), results, and conclusion (see Chapter 6). Figure 14.1 shows the traditional position of these components on a science fair display board.

Title

Your title may state the specific independent and dependent variables being investigated or may be worded creatively to capture the reader's interest.

- The Effect of Bay Leaves and Cucumbers on Cricket Behavior
- Repel Crickets with Cucumbers and Bay Leaves
- Methods of Warfare on Crickets

Statement of the Problem

You should clearly communicate the essential idea of your research through the problem statement. On display boards, the problem is frequently stated as a question followed by the specific hypotheses to be tested.

Problem: Will cucumber skins and bay leaves repel crickets?

Hypotheses:
1. Dried cucumber skins will repel crickets.
2. Dried bay leaves will repel crickets.
3. A combination of dried cucumber skins and bay leaves will be most effective in repelling crickets.

Procedure

The procedure (methods-materials) for your experiment may be communicated as lists or written in paragraph form. Because of limited space, the procedure must be briefly stated. Consult science fair guidelines for requirements.

> One hundred grams of bran were placed in one corner of a 20-gallon aquarium. Another 100 g of bran ringed with 25 g of dried bay leaves were placed in the opposite corner. Fifty crickets were placed in the aquarium. The distribution of crickets and the mass of bran consumed after 24 hrs. were recorded. Allowing a 1-day recovery period between each trial, the procedure was repeated 4 times. Similarly, the response of crickets to two piles

of bran alone (control), to cucumber skins surrounding one pile of bran, and to a combination of bay leaves and cucumber skins surrounding another pile of bran were determined.

Results

You should include sufficient data tables and graphs to communicate your findings and to show the extent to which the data support the research hypotheses. Typically, space does not permit the inclusion of numerous tables, graphs, or a lengthy discussion of the results and conclusions. Photographs or diagrams of experimental results are particularly effective. For example, the distribution of crickets could be depicted through photographs or diagrams, whereas the mass of bran consumed could be displayed through a graph. Brief sentences summarizing the data could accompany the diagrams and graph (see Figure 14.2). You must decide which display techniques best communicate your experiment. Supplementary data tables and graphs should be placed in the written report. Be sure to refer to them when you make an oral presentation to the judges.

Conclusion

In the conclusion section of a science fair display, you should summarize your major findings and the extent to which the results support the research hypotheses. Findings must be written concisely in paragraph form or as a list. A brief explanation of findings is also appropriate as part of the conclusion. Display space will not permit a lengthy discussion of results as in a written report. Major recommendations for additional research and improvements may be cited if space permits.

> Effective warfare against crickets can be waged with bay leaves but not with dried cucumber skins. Research data supported the

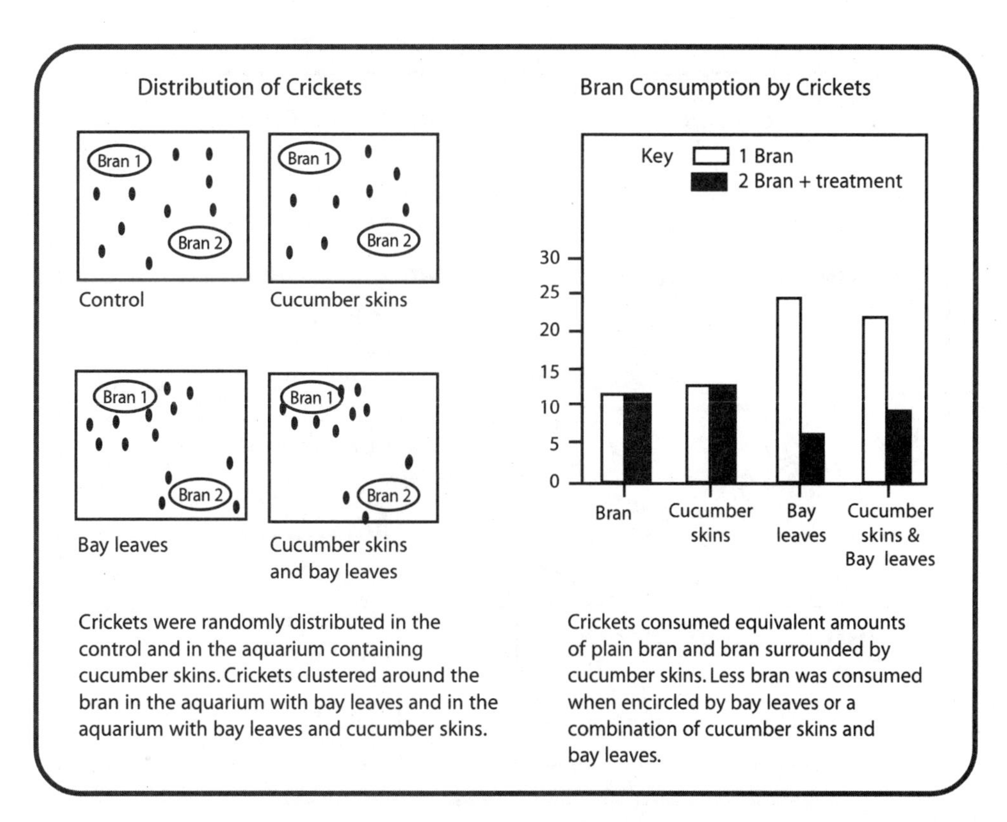

Crickets were randomly distributed in the control and in the aquarium containing cucumber skins. Crickets clustered around the bran in the aquarium with bay leaves and in the aquarium with bay leaves and cucumber skins.

Crickets consumed equivalent amounts of plain bran and bran surrounded by cucumber skins. Less bran was consumed when encircled by bay leaves or a combination of cucumber skins and bay leaves.

Figure 14.2 Data Displays for Science Fair Display, Methods of Warfare on Crickets.

hypothesized effectiveness of bay leaves but NOT the hypothesized effectiveness of cucumber skins NOR the superior repelling power of the combination. Future experiments should address the effectiveness of fresh materials and of mixing food and materials. Different cricket samples should be used to eliminate potential saturation or adaptation.

Constructing Display Boards

Numerous methods exist for constructing display boards. Appropriate materials include plywood, plexiglass, reinforced poster board, corrugated cardboard, and styrofoam sheets. The three-panel board may be hinged or reinforced with tape. Inexpensive project display boards are available in many large office supply stores. All lettering should be neat and legible; spelling should be carefully checked. Consider artistic appeal by selecting complementary colors that enhance photographs and display materials. Excellent publications on photography in science research are published by the Eastman Kodak Company. Plan the display carefully. Remember, it is the advertisement for the project. A sloppy display will not favorably impress judges and the public. See Table 14.2 *Science Project Evaluation Form* for a typical science project evaluation form used by judges.

POSTER DISPLAYS

In some schools, poster displays accompanied by written reports are used instead of science fair three-panel displays. In preparing a poster you can gain valuable experience in communicating your project in a small amount of space. A sample poster is shown in Figure 14.3.

ORAL PRESENTATIONS

Typical oral presentations required for many competitive events last approximately 10 minutes and follow the components of the formal research paper. Frequently, judges ask 5 minutes of ques-

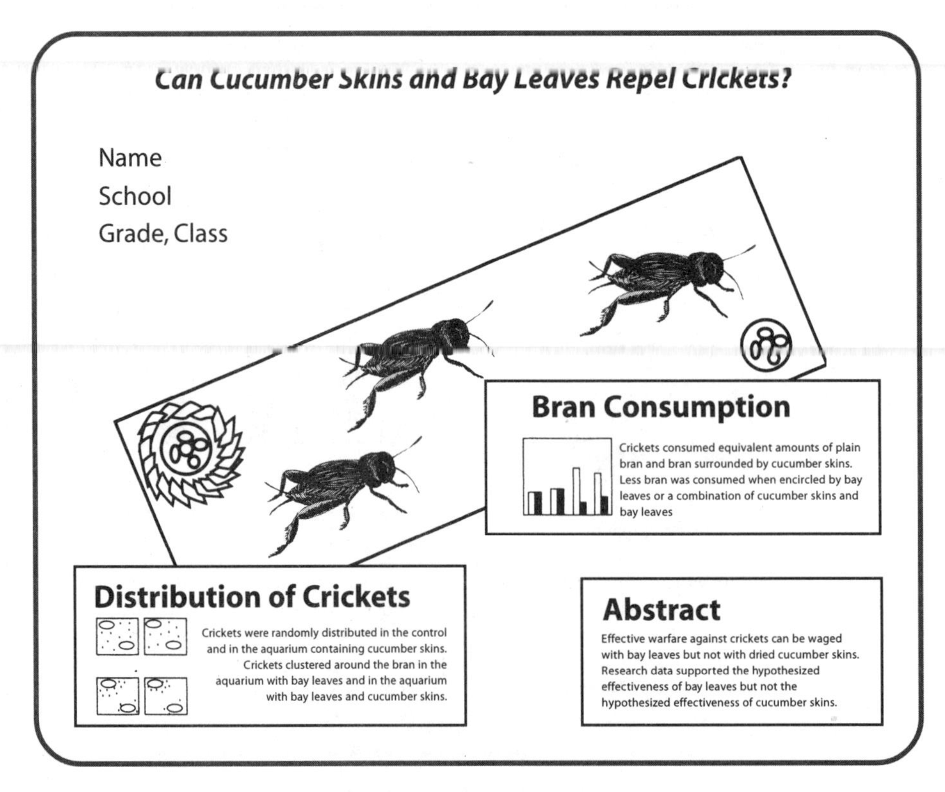

Figure 14.3 Poster Display for Science Project on Cricket Warfare.

tions after an oral presentation. When the format of a competition makes a display board or poster inappropriate, other visual aids, such as transparencies and slides, are critical to an effective presentation. Major components of the presentation and representative types of visual aids are outlined in Table 14.1 *Suggested Sequence and Visuals for Oral Presentations*.

You will be more comfortable with public speaking if you know and practice what will be expected. Use Table 14.1 or the following list of steps to help you plan your oral presentation.

- Begin by telling the audience about yourself and how you became interested in your project. Include your name, grade, and school.
- Describe your problem and give important background information on the variables in your study and in related research.
- State your purpose and research hypotheses.
- Describe the procedure you followed to test the hypotheses. Be sure to include information on your independent and dependent variables, the constants, the control, and the number of repeated trials. It is important that you also acknowledge any help you received in conducting the experiment.
- Explain your results using transparencies or slides of tables and graphs.
- Share your conclusions and state the extent to which they supported the original hypothesis.
- Suggest areas for future study and for improvement of the experiment.

You can change your written report into an oral presentation by using the following steps.

- Underline the most important facts in the research paper. Note the best graphs and tables. Describe visuals that may need to be added.
- Read the paper to a teacher, parent, or another student. Describe the visuals you will use. Ask for feedback on posture, speech, volume, inflection, speed, and proposed visuals.
- Write the important information on note cards. With a colored pen, designate visuals to be used.
- Prepare visuals.
- Using the visuals, read the note cards to a teacher, parent, or fellow student. Record the reading. Obtain feedback.
- Practice, practice, practice—in front of the mirror, at home, or with other students.
- Practice the presentation in front of a class. Obtain feedback.
- Continue to polish the presentation until you can speak (not read) from the note cards and can effectively integrate the visuals.

Finally, it is normal to be nervous, especially when you begin. Take a deep breath, stand straight, and begin. Create a good first impression by looking your best, by not chewing gum, and by maintaining eye contact with the judges and the audience. Be friendly and enthusiastic about your project. Enthusiasm is contagious. It can motivate the audience to be more attentive. Remember basic speaking principles include the importance of speaking clearly, at a reasonable pace, and loudly enough to be heard by everyone. Stand to the side of visual aids and point to the appropriate information on the visual aids. See Table 14.3 *Questions Frequently Asked by Judges* for a list of questions frequently asked by judges.

Related Web Sites

http://ibms50.scri.fsu.edu/~dennisl/CMS.html

http://134.121.112.29/sciforum/writing.html (Scientific Poster)

http://www.eduzone.com/Tips/science/showtip4.htm (Display Boards)

http://www.eduzone.com/Tips/science/SHOWTIP2.HTM (logistics for organizing science fair competition)

TABLE 14.1 Suggested Sequence and Visuals for Oral Presentations

Presentation Components	Representative Types of Visual Aids
Introduction	
1. Introduce yourself.	
2. Tell your audience how you became interested in the topic.	
3. State the problem.	■ Picture, diagram, table, or graph that illustrates the problem and sets the stage for the presentation
4. Review pertinent background information on variables and prior research.	■ Photographs or diagrams of experimental subjects or phenomena ■ Table summarizing similarities and differences among research studies
5. State the research hypotheses.	■ List of research hypotheses
Procedure (Methods-Materials)	
1. Describe the design of the experiment.	■ Diagram of the experimental design ■ List of materials
2. Describe the experimental procedures (methods and materials).	■ Diagrams or photographs of special apparatus ■ List of steps or a flow diagram ■ Photographs of experimental stages
3. Explain how the data were analyzed.	
Results	
1. Display the results.	■ Tables and graphs for each type of data
2. Describe the results.	■ Photographs of treatment groups
Discussion-Conclusion	
1. Summarize the major findings, including support for the research hypotheses.	■ List of major findings
2. Compare findings with prior research.	
3. Suggest improvements, topics for further study, and potential applications.	■ List of improvements, future research topics, applications

TABLE 14.2 Science Project Evaluation Form

Name ______________________ Number ______________

Please circle: 3 points for very strong; 2 points for moderately strong; 1 point for weak or nonexistent.

Criteria	Point Value		
Background Knowledge			
Key scientific concepts	3	2	1
Literature review	3	2	1
Experimental Design			
Hypothesis: Testable relationship between variables	3	2	1
Independent Variable (IV): Factor purposely changed	3	2	1
Dependent Variable (DV): Factor that responds	3	2	1
Constants: Factors kept the same	3	2	1
Control: Used as a standard for comparison	3	2	1
Repeated Trials: Number of subjects or times repeated	3	2	1
Procedure (Materials-Methods)			
All materials included	3	2	1
Clear, precise procedure	3	2	1
Results			
Data Tables: IV, DV, derived quantities, units	3	2	1
Graphs: correct type, scales, title, line-of-best-fit	3	2	1
Summary: sentences/paragraphs about tables/graphs; statement of how data supports hypothesis	3	2	1
Conclusion			
Major findings	3	2	1
Interpretation	3	2	1
Suggestions for further study	3	2	1
Significance/application	3	2	1
Display			
Attractive and legible	3	2	1
Accurate	3	2	1
Consistent with fair regulations	3	2	1
Interview			
Communicates scientific basis	3	2	1
Describes design principles	3	2	1
Explains data analysis	3	2	1
Recognizes limitations	3	2	1

TABLE 14.3 Questions Frequently Asked By Judges

Background Knowledge
Why did you decide on this topic? What is the purpose of your project? What library information did you find that was helpful?
Experimental Design
What was your hypothesis? What variable did you intentionally change? What response did you observe or measure? What did you intentionally keep the same? What group did you compare the others against? Why? How many times did you repeat the experiment?
Materials and Methods
What materials did you use? What steps did you follow in conducting the experiment? If you had a mentor, in what ways did the mentor assist you?
Results-Conclusion
What results did you find? How did your results relate to your original hypothesis? What conclusion did you make? If you conducted the experiment again, what would you do differently? What additional experiments would you suggest? Which groups in the community would be interested in your experiment? What recommendations would you make to these groups? What was the most important thing you learned from the experiment?

USING TECHNOLOGY

Presentation graphics programs

After you have written your report with a word processing program and used a calculator or spreadsheet to analyze your data, use a presentation graphics program to convince people you know what you are talking about. Use a program such as PowerPoint to create slides, outlines, and tables to share what you know. Visuals provide a framework for your presentation. You can talk about each visual and will not have to memorize a speech. In preparing for any competitive event, be sure to check the rules regarding presentation. Confirm that the kind of equipment you need is available, and always have a back up plan if a problem arises.

CHAPTER 15

Evaluating Your Knowledge of Experimental Design and Data Analysis

Want to see what you know? Answer the following questions as a self-test of your knowledge of experimental design and data analysis. When you have completed this test, turn to page 219 and check your answers. If you miss any questions, review the chapter indicated in parentheses at the beginning of each question.

1. (Chapters 1, 2) Match each term in Column II with its definition in Column I.

	Column I Definitions	Column II Terms
______	A. Used to reduce the effects of chance errors.	1. Independent variable
______	B. A statement of a possible relationship between the independent and dependent variables.	2. Repeated trials
______	C. The factor in an experiment that responds to the purposefully changed factor.	3. Control
______	D. A group or sample that is used as a standard for comparison.	4. Constant
______	E. The factor in an experiment that is changed on purpose.	5. Hypothesis
______	F. Any factor that is not allowed to change.	6. Dependent variable

2. (Chapters 1, 2) Read the following scenario of a soda-pop experiment. Use the information in the scenario to answer questions 2a–2g.

   Raheem wanted to determine if ice made soda-pop lose its fizz. He put 100 ml of "Molly's Root Beer" into each of 15 identical water glasses. To three of the glasses he added 1 ice cube each, to the next three he added 2 ice cubes each, to the next three he added 3 ice cubes each, and to the fourth set of three he added 4 ice cubes each. He used identical ice cubes. To a fifth set of 3 glasses he added no ice cubes. Raheem then measured the time it took for the root beer to go flat or stop releasing bubbles.

2a. What is the independent variable?
2b. What is the dependent variable?
2c. What are the constants?
2d. What is the probable hypothesis being tested?
2e. What levels of the independent variable are being tested?
2f. How many repeated trials were conducted?
2g. State a control for this experiment.

3. (Chapter 2) Diagram the experiment described in the following scenario. Be sure to include in your diagram the title and the hypothesis.

   Bobbie wanted to find out if diluting vinegar would change the speed at which it would react with Gray's Baking Soda. She used identical glasses. Into each of 3 glasses she put 2 g of baking soda and 25 ml pure vinegar; in each of 3 more glasses she put 2 g of baking soda, 20 ml of vinegar, and 5 ml water; into each of 3 more glasses she put 2 g of baking soda, 15 ml vinegar, and 10 ml water; into each of 3 more glasses she put 2 g of baking soda, 10 ml vinegar, and 15 ml water. In each case she measured the time it took the vinegar and the baking soda to finish reacting or to stop producing carbon dioxide gas bubbles.

4. (Chapter 2) What is the control, if any, in the scenario in Question 3?

5. (Chapter 2) List ways the experiment described in Question 3 could be improved.

6. (Chapter 3) The *Four Question Strategy* was designed to help you turn a general topic into an experimental design that includes all the major parts of an experiment. The four questions are:

   I. What materials are readily available for conducting experiments on ____________?
   (General Topic)

   II. How does ____________ act?
   (General Topic)

   III. How can I change the set of ____________ materials to affect the action?
   (General Topic)

   IV. How can I measure or describe the response of ____________ to the change?
   (General Topic)

   Answer each of the four questions for the general topic *Bird Seed.*

7. (Chapter 3) Choose two variables from Question 6 above and write a hypothesis.

8. (Chapter 3) Write a title for an experiment based on your hypothesis.

9. (Chapter 3) When you choose one of the responses to Question 6 of the *Four Question Strategy* as your independent variable, what becomes of the other responses in the list?

10. (Chapter 5) A description of an investigation and a table of data are given below. Draw a grid like the one shown below on your paper and use it to: a) construct a graph of the data; b) draw a best-fit line; and c) write a description of the relationship between the variables.

Tomato seeds were kept moist and exposed to different temperatures to see if the number that had germinated after two weeks would vary. The following data were obtained:

Temperature (°C)	Average number of seeds that germinated
15	20
20	47
25	87
30	74
35	6
40	0

11. (Chapter 5) The graph below shows data collected while watching house flies. Construct a data table for the graph.

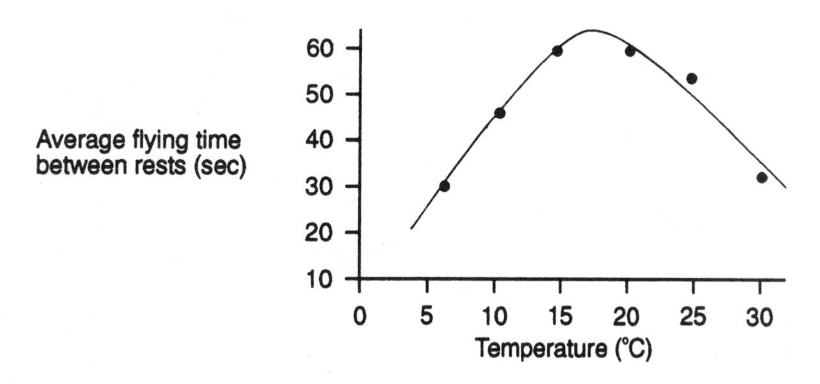

12. (Chapter 5) When plotting qualitative (discrete) data, you should draw:
    a. a bar graph
    b. a line graph
    c. either a bar or a line graph
    d. neither, qualitative data cannot be graphed.

13. (Chapter 5) When plotting quantitative (continuous) data, you should draw:
    a. a bar graph
    b. a line graph
    c. either a bar or a line graph
    d. none of the above

14. (Chapter 5) Use the following data table on color of shirts worn on specific holidays to answer the question.

Color Shirts	Holiday		
	Christmas	St. Patrick's	4th of July
Red	5	2	4
Blue	1	1	4
Green	4	6	1
Orange	0	4	0
White	3	0	4

This data could be acceptably plotted as:

a. a frequency distribution
b. a line graph
c. neither a or b
d. either a or b

15. (Chapter 5) Use the following data table about the effects of phosphate concentration on algae growth to answer this question.

Phosphates (%)	Amount of algae (g)
2	6
4	12
6	18
8	24

This data could be acceptably plotted as:

a. a bar graph
b. a line graph
c. neither a or b
d. both a and b

16. (Chapter 5) For each experiment title listed below, state whether the experiment should be graphed as a *bar* or a *line* graph.

a. The Effect of an Increase in the Temperature on the Number of People at Jackson Beach.
b. The Effect of Gender on the Color of Notebook Purchased.
c. The Effect of the Amount of Sugar in the Solution in a Hummingbird Feeder on the Number of Hummingbird Visits.
d. The Effect of the Number of Hunters in an Area on the Number of Deer in the Same Area After the Hunting Season.
e. The Effect of the Color of Sunglasses on the Number Sold.

17. (Chapter 8) Use the following experimental situation to answer questions 12a to 12g. Aaron raised moles. From one litter he obtained moles with the following masses:

Mole	Mass (g)
1	4
2	10
3	6
4	4
5	10
6	11
7	4

17a. Which type of data is represented (qualitative or quantitative)?

17b. What is the *minimum value* of the masses of the moles?

17c. What is the *range* in mole mass?

17d. Compute the mean of the data.

17e. Compute the mode of the data.

17f. Compute the median of the data.

18. (Chapter 8) List 3 measures of central tendency:

a.

b.

c.

19. (Chapter 8) List 3 measures of variation within a set of data:

a.

b.

c.

20. (Chapter 8) Megan wanted to determine if temperature had an effect on the length of time it took a certain species of mealworms to mature from the worm stage to the adult stage. She kept 5 mealworms at each of 3 temperatures and recorded the number of days it took for the worms to become adults. At 18°C it took 22, 22, 24, 21, and 21 days. At 22°C it took 18, 16, 19, 17, and 20 days. At 26°C it took 14, 16, 13, 14, and 13 days.

a. What kind of data is represented by the dependent variable?

b. What level of measurement is represented by the dependent variable?

c. What is the most appropriate measure of central tendency?

d. What is the most appropriate measure of variation?

e. Construct a data table for summarizing the data.

f. Compute the appropriate measures of central tendency, variation, and number; enter them in the table.

21. (Chapter 8) Derk wondered if the color of the wrapper on a piece of hard candy would affect how often it was selected by kindergartners. The same flavor of candy was wrapped in red, blue, green, brown, and orange colored wrappers. For the month of February, he collected the following data:

	Red	Blue	Green	Brown	Orange
Ms. Silverstein's class	29	31	12	10	18
Mr. Jackson's class	22	19	16	30	13
Ms. Bealton's class	16	32	17	18	17

a. What type of data is represented by the dependent variable?
b. What level of measurement is represented by the dependent variable?
c. What is the most appropriate measure of central tendency?
d. What is the most appropriate measure of variation?
e. Construct a table for summarizing the data.
f. Compute the appropriate measures of tendency, variation, and number; enter them in the table.

USE THE FOLLOWING SCENARIO TO ANSWER QUESTIONS 22-33.

In Physics class, Alicen and Jennifer were studying traction. They conducted an experiment titled, "The effect of the surface of an incline on the time it takes a battery-operated toy truck to pull two #2 sinkers 100 cm up a 7% grade." Their data were:

Medium Sand Paper			Ultrafine Sand Paper			Varnished Wood			Formica		
10	14	15	13	16	13	13	10	14	23	22	18
9	12	11	16	15	15	14	16	16	21	21	21
16	14	11	14	13	17	16	17	14	19	23	24
14	15	14	15	14	14	14	15	15	24	20	20
12	14	15	17	17	16	17	17	13	21	19	21
11	12	13	13	14	17	15	11	17	21	21	22
12	13	14	15	13	14	16	17	16	22	22	20
15	16	15	14	16	13	15	12	13	20	20	23
11	18	11	16	15	15	17	17	17	20	22	18
20	14	10	17	17	16	16	12	17	19	21	22

22. (Chapter 10) Complete the following table:

	Min	Max	Q_1	Median	Q_3	Range
Medium Sandpaper						
Ultra-fine Sandpaper						

23. (Chapter 10) Which set of data for either the medium sandpaper or the ultra-fine sandpaper is skewed?

24. (Chapter 10) How can you tell?

25. (Chapter 10) Does the data for the medium sandpaper or the ultra-fine sandpaper have more variation in it? How can you tell?

26. (Chapter 10) For which (if any) of the surfaces is the distribution a J-distribution?
   a. medium sandpaper
   b. ultra-fine sandpaper
   c. varnished wood
   d. Formica
   e. none of the above

27. (Chapter 10) For which (if any) of the surfaces is the distribution a U-distribution?
   a. medium sandpaper
   b. ultra-fine sandpaper
   c. varnished wood
   d. Formica
   e. none of the above

28. (Chapter 10) For which (if any) of the surfaces is the distribution a rectangular distribution?
   a. medium sandpaper
   b. ultra-fine sandpaper
   c. varnished wood
   d. Formica
   e. none of the above

29. (Chapter 10) For which (if any) of the surfaces is the distribution a normal distribution?
   a. medium sandpaper
   b. ultra-fine sandpaper
   c. varnished wood
   d. Formica
   e. none of the above

30. (Chapter 10) What is the standard deviation for the varnished wood data?

31. (Chapter 10) What is the standard deviation for the Formica data?

32. (Chapter 11) Using a t test, determine if the difference between the data for the ultra-fine sandpaper and the varnished wood is significant? What is the t value? What is the probability of the t value?

33. (Chapter 11) Using a t test, determine if the difference between the data for the medium sandpaper and the ultra-fine sandpaper? What is the t value? What is the probability of the t value?

34. (Chapter 10) In Question 20, are there statistically significant differences in the lengths of time it took mealworms kept at different temperatures to mature?

35. a. (Chapter 10) In Question 21, is the distribution of the choice of color of the wrapper in Ms. Silverstein's class significant? Show all your work.
    b. Which color(s) least influenced the results?

ANSWERS TO CHAPTER 15 SELF-TEST OF YOUR KNOWLEDGE OF EXPERIMENTAL DESIGN AND DATA ANALYSIS

1. A 2
   B 5
   C 6
   D 3
   E 1
   F 4

2. a. The number of ice cubes
   b. The time it takes the soda to go flat or stop releasing bubbles
   c. Water glasses, brand of root beer, amount of root beer, the ice cubes—size, shape, composition.
   d. If the amount of ice in a soda is increased, then it will lose its fizz faster (or slower).
   e. 0, 1, 2, 3, ice cubes
   f. 3
   g. The 0 ice cube set is a no treatment control

3.

Title: The Effect of the Concentration of a Vinegar Solution on the Rate of its Reaction with Baking Soda
Hypothesis: If the concentration for a vinegar solution is decreased, then it will react with baking soda more slowly.

IV: Concentration of vinegar			
Pure 25 ml vinegar (control)	20 ml vinegar/ 5 ml H_2O	15 ml vinegar/ 10 ml H_2O	10 ml vinegar/ 15 ml H_2O
3 Trials	3 Trials	3 Trials	3 Trials

DV: Time it takes for the reaction to stop
C: Glasses
Amount of baking soda

4. Pure vinegar is a control since it is undiluted. You could also have used 25 ml of water and 0 ml of vinegar as a control.

5. Specify the kind of vinegar to be used throughout.
   Specify the kind of baking soda to be used throughout.
   Specify the temperature of the water, vinegar, and baking soda.

6. Question 1.
   Feeders
   Seeds

   Question 2.
   Seeds are eaten.
   Seeds are scattered.
   Seeds are selected.

   Question 3.

Feeders	**Seeds**
Size	Mixes
Shape	Single type
Color	Color
Height	Size
Location	Alternate food sources
Surroundings	
Dispensing system	
Perch area	

   Question 4.
   Measure the rates at which different seeds are eaten.
   Measure amounts of different seeds left.
   Kind of seeds dropped to the ground.
   Number of different seeds dropped to the ground.
   Total amount of seeds eaten.
   Number of birds coming to the feeder.
   Types of birds coming to the feeder.

7. Examples

   If the height at which bird feeders are hung is increased, then the amount of seeds eaten will increase.

   or

   If a mixture of seeds is placed in a feeder, then the sunflower seeds will be eaten the fastest, the corn seeds will be eaten second fastest; and the millet will be eaten least.

8. The Effect of the Height of a Birdfeeder Above the Ground on the Amount of Seed Eaten

   or

   The Effect of the Kind of Seed on the Eating Rate by Birds

9. They become constants.

10. The Effect of Temperature on the Number of Seeds that Germinate

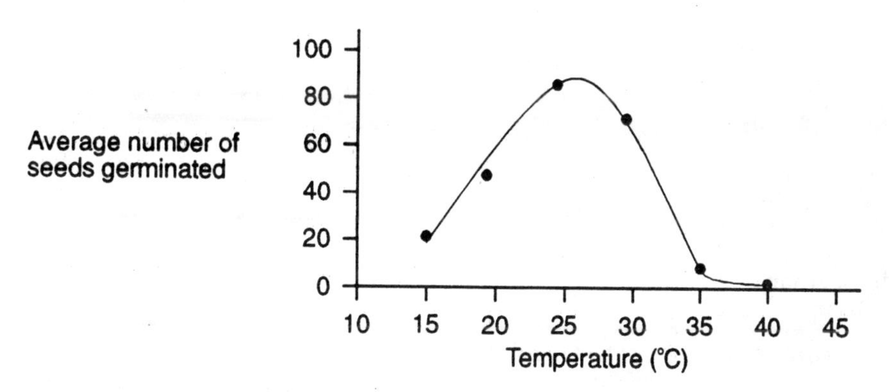

C. As the temperatures increased from 15°C to 25°C the number of seeds that germinated increased, but above 25°C the number of seeds that germinated decreased rapidly.

11. The Effect of Temperatures (°C) on the Flying Time of House Flies.

Temperatures (°C)	Average flight time (sec)
5	30
10	45
15	60
20	57
25	55
30	30

12. a.

13. c.

14. a.

15. d.

16. a. line or bar graph
    b. bar graph
    c. line or bar graph
    d. line or bar graph
    e. bar graph

17. a. quantitative
    b. 4g
    c. 7g
    d. 7g
    e. 4g
    f. 6g

18. mean, mode, median

19. Frequency distribution, range, standard deviation

20. a. quantitative
    b. ratio
    c. mean
    d. range or standard deviation
    e. & f. The Effect of Temperature on the Number of Days It Takes Mealworms to Mature.

	Temperature (°C)		
Descriptive information	**18**	**22**	**26**
Mean	22	18	14
Range	3	4	3
Max	24	20	16
Min	21	16	13
Number	5	5	5

21. a. Qualitative
    b. Nominal
    c. Mode
    d. Frequency distribution
    e. & f. The Effect of the Wrapper Color on Its Selection by Kindergarten Students.

	Color preferred		
Descriptive information	**Silverstein's class**	**Jackson's class**	**Bealton's class**
Mode	**Blue**	**Brown**	**Blue**
Frequency distribution			
Red	29	22	16
Blue	31	19	32
Green	12	16	17
Brown	10	30	18
Orange	18	13	17
Number	100	100	100

22.

	Minimum	Maximum	Q_1	Median	Q_3	Range
Medium Sandpaper	9	19	11	14	15	10
Ultra-fine Sandpaper	13	17	14	15	16	4

23. Medium Sandpaper

24. The median is much closer to Q_3 than to Q_1 for the medium sandpaper. The median is mid-way between Q_3 and Q_1 for the ultra-fine sandpaper.

25. The medium sandpaper has the greater variation because it has the greater range and a greater difference between Q_1 and Q_3.

26. c

27. a

28. b

29. D

30. 2.45

31. 1.44

32. No; t = .07391; p = .9413

33. Yes; t = 3.025; p = .00370

34.

$$S^2_{18°C} = \frac{\Sigma(x_i - \overline{x})^2}{N-1}$$

$$= \frac{(22-22)^2 + (22-22)^2 + (24-22)^2 + (21-22)^2 + (21-22)^2}{(5-1)}$$

$$= \frac{(0)^2 + (0)^2 + (2)^2 + (-1)^2 + (-1)^2}{4}$$

$$= \frac{6}{4}$$

$$= 1.5$$

$$S^2_{22°C} = 2.5$$

$$S^2_{26°C} = 1.5$$

$$t = \frac{\overline{X}_1 - \overline{X}_2}{\sqrt{\frac{S_1^2 + S_2^2}{n}}}$$

$$t_{18°C \text{ versus } 22°C} = \frac{22 - 18}{\sqrt{\frac{1.5 + 2.5}{5}}}$$

$= 4.5$ $\quad dt = 8 \quad p < .01$

$t_{18°C \text{ versus } 26°C} = 10.4 \quad df = 8 \quad p < .01$

$t_{22°C \text{ versus } 26°C} = 4.5 \quad df = 8 \quad p < .01$

35a. $$\chi^2 = \sum \frac{(O - E)^2}{E}$$

$$= \frac{(31-20)^2}{20} + \frac{(29-20)^2}{20} + \frac{(18-20)^2}{20} + \frac{(12-20)^2}{20} + \frac{(10-20)^2}{20}$$

$$= \frac{(11)^2}{20} + \frac{(9)^2}{20} + \frac{(-2)^2}{20} + \frac{(-8)^2}{20} + \frac{(-10)^2}{20}$$

$$= 6.05 + 4.05 + .2 + 3.2 + 5$$

$= 18.5 \quad df = 4$

$= 18.5 > 14.86 \quad p < .001$

35b. Orange

Glossary

abstract—a brief summary of a research paper or report. It includes a brief description of the problem, hypothesis, procedures, results, and conclusions.

average (mean)—the most central or typical value in a set of quantitative data; the formula for calculating the mean is:

$$mean = \frac{\text{the sum of all the measurements (or counts)}}{\text{total number of measurements (or counts)}}.$$

axis—the horizontal or vertical line found at the bottom and left side of a graph; plural is axes.

back to back stem and leaf plots—a plot of two sets of data that are compared by having the leaves extend in opposite directions from the same stem.

bar graph—a pictorial display of a set of data using bars to indicate the value, amount, or size of the dependent variable for each level of the independent variable tested. Usually the taller the bar, the greater the value of the dependent variable.

bias—a statistical error that occurs when samples are drawn so that all members of the population do not have an equal chance of being included.

bibliography—a list of all books, papers, journal articles, and communications cited or used in the preparation of a report or scientific research paper.

box and whisker plot—an exploratory plot of a data set that displays the median and information about the range and distribution or variance of the data such as the minimum, maximum, and each quartile.

call number—the set of numbers and letters used in a library to identify, catalog, and shelve each book.

central tendency—the value that is most typical of a set of data.

chi-square test—a test for qualitative data used to determine if differences between frequency distributions are statistically significant.

cluster—a subset of data within a data set, whose values are very close together.

conclusion—the last section of a report of an experiment; it states the purpose, major findings, hypothesis, a comparison of the findings of this experiment with other experiments and recommendations for further study.

constants—those factors in an experiment that are kept the same and not allowed to change or vary.

continuous data—measurements made using standard measurements scales that are divisible into partial units. Time and volume are examples of continuous data. When both the independent and dependent variables result in continuous data, the data can be graphed as *either* a line or a bar graph.

control—the part of an experiment that serves as a standard of comparison. A control is used to detect the effects of factors that should be kept constant, but which vary. The control may be a "no treatment" group or an "experimenter selected" control.

counts—data stating the number of items, for example, the number of bees attracted to sugar water or the number of people responding to a noise.

data—the bits of information (measurements, observations, or counts) gathered in an experiment; data takes a plural form verb, "the data *are;* the data *were*".

data table—a chart to organize and display the data collected in an experiment.

degrees of freedom—the number of independent observations in a sample. In a sample of n members with a fixed mean, the degrees of freedom are (n – 1).

dependent variable (responding variable)—the factor or variable that may change as a result of changes purposely made in the independent variable.

derived quantity—information or values determined by calculations using collected data; examples are means, medians, modes, and ranges.

descriptive statistics (summary statistics)—statistics that describe for a set of data the most typical values and the variations that exists within the set.

discrete (discontinuous) data—data that exists in categories that are separate and do not overlap such as brands of products and kinds of papers. When displayed by a scale on a graph, the points between the defined categories do not have any meaning. Discrete data can be graphed as a bar graph but *not* a line graph.

error (in measurement)—the unavoidable chance errors that can result in increasing or decreasing the value of individual measurements.

experiment—a test of a hypothesis. It determines if purposely changing the independent variable does indeed change the dependent variable as predicted.

experimental design diagram—a graphic illustration of an experiment

Title:
Hypothesis:

IV:				

independent variable

levels of the IV tested and control

number of repeated trials

DV: dependent variables
C: constants

experimenter selected control—the set of trials conducted for a single level of the independent variable that is selected by the experimenter to be the standard of comparison. For example, in an experiment to determine which brand of gasoline is best a "no treatment" control, using no gasoline in a test car does not make sense.

Four Question Strategy—an approach for generating a series of experiments from a topic, demonstration, or other prompt. Brainstorming responses to the four questions in the strategy results in many potential independent variables and dependent variables to select among. It also results in descriptions of ways to describe or measure the dependent variable and identify many constants.

frequency distribution—a summary or graph of the amount of variation (spread) within a set of qualitative data (observations); a frequency distribution states the number of items in each category, for example, 2 red, 10 pink, and 25 green tomatoes.

gap—data values within a data set for which there are no entries.

graph—a pictorial display of a set of data.

horizontal (X) axis—is the line along the bottom of a graph on which the scale for the independent variable is placed.

hypothesis—a prediction of the relationship of an independent and dependent variable to be tested in an experiment; it predicts the effect that the changes purposely made in the independent variable will have on the dependent variable. Plural is hypotheses.

independent variable (manipulated variable)—the variable that is changed on purpose by the experimenter.

inferential statistics—the methods or mathematical procedures by which to determine if the variations between sets of data are statistically significant or not. The statistics that indicate the likelihood that the variation occurred by chance or as a result of the treatments.

interquartile range—the difference between data values of the upper and lower quartiles ($Q_3 - Q_1$).

interval data—measurements made using a scale with equal intervals, but no absolute zero. For example °C.

intervals—the equal size values represented by the equal spaces marked along the axis of a graph, or the spaces between units of a measuring device.

introduction—a paragraph at the beginning of a report of an experiment that states why the experiment was done (reason); what was expected to be learned (purpose) by doing it, and the hypothesis tested.

J-shaped distribution—a distribution of data in which the value of the data that occurs at the greatest frequency is at the extreme of one end of the data values, indicating a probable limit of values.

level of significance—the level of probability set by the experimenter for rejecting the null hypothesis. It is the level of probability that the experimental results were due to the treatment and not to chance.

levels of the independent variable—the specific values (kinds, sizes, or amounts) of the independent variable that are tested in an experiment.

line graph—a pictorial display of data that can be drawn when the data for both variables are continuous data. The line in a line graph shows the relationship between the independent and dependent variables.

line of best fit—a smooth line drawn so that the totals of the distances between the line and the points above and below it are equal (roughly half the data points are above and half are below the line). The line that can be either a straight line or a smooth curve shows the relationship between the independent and dependent variable.

lower extreme—the minimum data value in a data set.

lower quartile (Q_1)—the number (data value) below which 25% of the values in a data set fall.

manipulated variable—see independent variable.

mean—see average.

measurements—data collected using a measuring instrument with a standard scale.

median—the central value in a set of data ranked from highest to lowest. Half the data are above it and half are below.

middle quartile (Q_2)—the mean, the data value below and above which 50% of the values in a data set fall.

mode—the most typical or central value of a set of qualitative data. It is the value that occurs most often in the set.

negative association—the inverse relationship that exists when increasing the independent variable results in a decrease of the dependent variable.

nominal data—data for a series of discrete categories for which there is not a basis for rank ordering, for example gender or hair color.

no treatment control—a control that receives none of the independent variable, for example in an experiment testing the effect of varying the amount of a fertilizer on plant growth, a no treatment control would be a set of plants that receives no fertilizer.

note card—a card used to record notes about the information found in a reference source or interview. Reference documentation is also put on each note card.

normal (bell shaped) distribution—a distribution of data in which the value of the mean is the most frequent data value and at each distance above and below the mean there is an equal frequency of data values.

null hypothesis—a hypothesis based on the assumption that two samples are from the same population and therefore have identical means or means which differ no more than would be expected by chance.

observations—data that are descriptions of qualities such as shape, color, and gender.

ordered stem and leaf plot—is a stem and leaf plot in which the data has been entered in rank order usually from the smallest value to the largest.

ordinal data—data collected for categories that can be rank ordered.

outliers—data values $< Q_1 - 1.5(Q_3 - Q_1)$ or data values $> Q_3 + 1.5(Q_3 - Q_1)$. These data values are considered to be unreasonable.

population—all members (persons or things) of a specific group that share a set of common characteristics.

positive association—the relationship that exists when increasing the independent variable results in an increase in the dependent variable.

probability (p) of error—is equal to the level of significance when the calculated value of a statistic and the table values are equal.

procedure—a sequence of precisely stated steps that describe how an experiment was done, including the materials and equipment used.

qualitative data—verbal descriptions or information gathered using scales without equal intervals or zero points. Such scales are non-standard scales.

quantitative data—information (data) gathered from counts or measurements using scales having equal sized intervals and a zero value. Such scales are standard scales.

range—a measure of how a set of measurements or count data is spread out. It is calculated by subtracting the minimum value from the maximum value.

ratio data—quantitative data collected using a scale with equal intervals and an absolute zero.

reference documentation—the information needed to identify each source used in doing and reporting an experiment; this includes the author's name, the titles of the articles and or books, newspapers, or journals, as well as the date, city and state of publication, and the publisher; or, in the case of an interview, the name of the person interviewed and the time, date, and place of the interview.

reference style manual—a book that states rules for writing the reference information for the books, interviews, encyclopedias, magazines, journals, and newspaper articles used in doing and reporting an experiment.

repeated trials—the number of times that a level of the independent variable is tested in an experiment or the number of objects or organisms tested at each level of the independent variable.
responding variable—see dependent variable.
results—a section of the report of an experiment that includes the data tables, graphs, and sentences that summarize any trends found in the data.
sample—the specific set of individuals, selected from a population, to be the subjects in an experiment.
sample bias—a sampling error that occurs when samples are drawn so that all members of a population do not have an equal chance of being included.
scale—a series of equal intervals and values placed on each axis of a graph.
skewed data sets—are data sets in which there are more data values above or below the mean.
standard deviation—the square root of the variance. It is a measure of how closely the individual data points cluster around the mean.
standard error of the mean—the statistic that indicates the extent to which sample means from the same population are expected to differ.
standard scale—a scale of measurement that has both a defined zero point and equal intervals. Examples are distance scales in inches, centimeters; temperature scales in degrees Celsius or Fahrenheit.
statistic—a number used to describe or analyze a set of data.
statistics—the branch of mathematics involved in collecting, analyzing, and interpreting data.
statistical significance—the criterion for the decision that the results of an experiment did not happen by chance but were the result of the treatment.
stem and leaf plot—an exploratory means of plotting data based on place value. The greatest place value, such as the tens or hundreds, is the stem, and the smaller place value, such as the ones, is the leaf. This plot shows the distribution of the raw data.
***t* test**—an inferential statistical test that is used to determine whether significant differences exist between the means of two samples.
title—a statement describing an experiment or data table. Titles are often written in the form, "The Effect of Changes in the Independent Variable on the Dependent Variable." (In a title all words containing four or more letters are capitalized.)
trend—the general direction or pattern of the data; it is usually illustrated on a graph as a line of best fit.
types of data—the kinds of information collected in an experiment; typical types are measurements, counts, or observations.
U-shaped distribution—a distribution in which two values of the data occur in higher frequencies than the other data values.
unordered stem and leaf plot—a stem and leaf plot in which the data is entered in an unranked order.
upper quartile (Q_3)—the number (data value) below which 75% of the values in a data set fall.
upper extreme—the maximum data value in a data set.
value—the size, amount, or extent of a property described by a piece of data, count, or observation.
variable—things or factors that can be assigned or take on different values in an experiment.
variance—the square of the standard deviation or the average squared distance from the mean.
variation—statistics that describe how spread out the values in a set of data are.
vertical (Y) axis—the line drawn on the left side of a graph on which the scale for the dependent variable is placed.

Appendix A: Using Technology

This appendix contains four resources:

I. Steps for graphing data on Texas Instrument's TI-83 graphing calculator.
II. Steps for graphing data on Casio's 9850 G+ graphing calculator.
III. TI-83 calculator program for a one-way Chi-square analysis. (See Chapter 11)
IV. Casio G+ calculator program for one-way Chi-square analysis. (See Chapter 11)

Note: There are often multiple ways to accomplish the same result on a calculator. The following steps are but one way. For other ways, consult the manual that accompanied the calculator.

I. Steps for graphing data on Texas Instrument's TI-83 graphing calculator.

1. Press the **ON** key

2. *Check the MODE* (setting how numeric entries are interpreted and displayed)
   - Press **MODE** key, check that all functions are highlighted to the left.
   - If not, arrow down to highlight and change each line as needed and press **ENTER** for each change.

3. *Clear lists* (removing old data)
   - Press the **STAT** key.
   - Press a **1** on the keyboard (for 1:edit) or simply press **ENTER**.
   - If the lists have old data:
     - Arrow to the top of a list to highlight.
     - Press **CLEAR** and then, **ENTER**.
     - Repeat for each list.

4. *Clear Y* = (removing old equations)
   - Press the **'Y='** key.
   - Move blinking cursor to the **'Y='** to be cleared.
   - Press **CLEAR** to remove old equation.
   - Arrow down to the other **'Y='** and press **CLEAR** to remove.

5. *Set the STAT PLOT* (defining how to plot statistical data)
   - Press **2nd** key and then the **'Y='** key.
   - To define plot 1, press **ENTER**.
   - Use arrows to highlight desired selections, press **ENTER** after each selection.
   - Selections are:
     - On or off
     - Type of graph (scatter, line, bar, box-and-whisker)
     - X list for independent variable
     - Y list for dependent variable
     - Type of mark (squares, crosses, dots)

- To define additional plots, arrow up and across to highlight plots 1, 2, or 3 and press **ENTER.**
- Use arrows to highlight selected plot number and press **ENTER.**
- Highlight desired selections as previously described.
- To turn on or turn off *all* plots, highlight numbers 4 or 5, press **ENTER.**

6. *Set window* (defining the viewing window, i.e., setting the interval scales for the axes)
   - To do this automatically, press **ZOOM** and then **9.**
   - To manually determine the boundaries and other attributes of the viewing window, see 'defining the viewing window' of the TI-83 guidebook.

7. *Graphing the data* (displaying the data)
   - Press **ZOOM 9** and you should see a graph of your data points.
   - Or set window manually and press the **GRAPH** key.

8. *To draw a line-of-best fit, use a mathematics technique called regression*

   If the data looks straight, try a linear regression analysis.
   If the data looks curved, try a quadratic analysis or exponential or power regression.

   - Press the **STAT** key, and arrow over to **CALC**; then arrow down to your choice, e.g., Linear regression [Lin Reg (ax+b)] and press **ENTER.**
   - Your choice will appear on the screen, e.g., LinReg (ax+b).
   - Tell the calculator where the data are, e.g., if the data are in lists 1 and 2, press **L1** (press 2nd key and 1), then press **comma**, then **L2** (press 2nd key and 2). Your screen should look like: LinReg (ax+b) L1,L2
   - To copy the resulting equation to a Y=, follow the steps below:
     - Place a comma after L2 so that the screen looks like: LinReg (ax+b) L1,L2,
     - Press the **VARS** key, and arrow over to **Y-VARS.**
     - Press **ENTER** or **1** to select **FUNCTION.**
     - On the **FUNCTION** menu, select a Y= (e.g., Y1) and press **ENTER.** Your screen should now look like: LinReg (ax+b) L1,L2,Y1 and a blinking cursor. Press **ENTER** and the calculator will calculate the equation and will also paste a copy in the designated Y=. That equation in 'Y=' will allow you to superimpose a line-of-best fit over your data points.
     - Press the **GRAPH** key and a line-of-best fit should appear on your graph.
     - Press the **Trace** key to trace the points of the graph. To trace the line-of-best fit, press the 'up' arrow to shift the cursor to the equation line-of-best fit. Moving the cursor along the line will allow you to predict values not directly measured.

II. Steps for graphing data on Casio's 9850 G+ graphing calculator.

1. *Access the MAIN MENU*
   - Press the **MENU** key.

2. *Choose the STAT icon*
   - Use the circle of arrow keys to highlight **STAT** (for statistics), press the blue **EXE** key.
   - Or simply press #2 on the keypad.

3. *Set the window to automatic*
   - To automatically have the calculator select the appropriate window (interval scales) for the graph, press **SHIFT,** then **MENU** to access SETUP.
   - 'Stat Wind' refers to statistics window, which you can set as automatic or manual.
   - Note the row of **'F'** keys below the screen. These keys control what appears along the bottom of the screen. Press **F1** to select automatic if necessary.
   - Press the **EXIT** key to return to the lists.

4. *Delete old data from lists*
   - Use the **F6** key to access a screen-bottom menu that includes **DEL-A** [for delete all].
   - To delete old data in List 1, for example, use the arrow keys to highlight anywhere in List 1 and press **F4, DEL-A** [for delete all]; press **F1** [for yes].
   - Arrow over to List 2 and repeat steps as necessary to delete data.

5. *Enter independent variable data in list 1* (e.g., drop height data)
   - Press **EXE** after each entry.

6. Enter dependent variable data in List 2 (e.g., bounce height data); check that you have correctly entered all data.

7. *Begin to graph the data*
   - Press **F6** to access a screen-bottom menu that has **GRPH** [for graph] above **F1**; press **F1.**

8. *Set up StatGraph 1 for a scatter plot*
   - Press **F6** [for SET] to make graph choices.
   - Arrow down to *Graph Type* and use the **F** keys to select **F1, Scat** [for scatter plot].
   - Arrow down to make other choices: XList (1), YList (2), frequency (1), mark type, and color.
   - To simultaneously set up additional graphs, arrow back to the top and highlight StatGraph 1. Use the **F** keys to select GPH2 or GPH3.
   - **EXIT** out when finished setting up the graph by pressing the **EXIT** key.

9. *Graph the data*
   - From this bottom menu, press **F1** to see the graph [GPH1].
   - If you have set-up two or three graphs, press **F4** [for SEL select] and use the **F** keys to turn *on* or *off* Graphs 1, 2, and 3 as desired. Then press **F6** [to DRAW the graphs].

10. *Draw a line-of-best fit on a scatter plot (so you can see the general trend of the data)*
    - From the bottom menu, press **F1** [for x, linear regression] for example.
    - From this screen, press **F5** to copy the equation to an empty 'Y=', press [**EXE**] to store.
    - Then press **F6** to draw a line over the scatter plot. See Step 11 to trace this equation line.

11. *To trace a line-of-best fit (so you can predict other values)*
    - After completing STEP 10, press **MENU** and select Graph.
    - Press **F6** (DRAW) to view the equation line-of-best fit.
    - Press **SHIFT F1** to access the Trace Function.
    - Use the left and right arrows to move the cursor along the line to predict new value for y.

Note: To display other types of graphs, select different graph types in *Step 8.*

III. TI-83 calculator program for one-way (one-independent variable) Chi-square analysis. (See Chapter 11)

```
ClrList (L1)
ClrHome
Disp "ENTER NUMBER"
Disp "OF CLASSES"
Input N
For(I,1,N,1)
ClrHome
Disp "CLASS"
Disp I
Disp "NUMBER OBSERVED"
Input O
Disp "NUMBER EXPECTED"
Input E
(O-E)²/E→L1(I)
End
ClrHome
Disp "CHI SQUARE VALUE"
Disp ""
Disp sum(L1)
Disp ""
Disp ""
```

IV. Casio G+ calculator program for one-way (one-independent variable) Chi-square analysis. (See Chapter 11)

```
"ENTER NUMBER"↵
"OF CLASSES"?→N↵
N→Dim List 1↵
For 1→I To N Step 1↵
ClrText↵
"PRESS [EXE] TO"↵
"ENTER DATA FOR"↵
"CLASS":I◢
"NUMBER OBSERVED"?→O↵
"NUMBER EXPECTED"?→E↵
((O-E)²)÷E→A↵
A→List 1[I]↵
Next↵
ClrText↵
"CHI SQUARE VALUE"↵
""↵
(Sum List 1)◢
""↵
```